Applied Mathematical Sciences
Volume 76

T0135032

Applied Mathematical Sciences

(continued in back)

P.A. Lagerstrom

Matched Asymptotic Expansions

Ideas and Techniques

Springer-Verlag
New York Berlin Heidelberg
London Paris Tokyo

P.A. Lagerstrom
California Institute of Technology
Applied Mathematics 217-50
Firestone Laboratory
Pasadena, CA 91125, U.S.A.

Editors

F. John
Courant Institute of
  Mathematical Sciences
New York University
New York, NY 10012
U.S.A.

J.E. Marsden
Department of
  Mathematics
University of California
Berkeley, CA 94720
U.S.A.

L. Sirovich
Division of
  Applied Mathematics
Brown University
Providence, RI 02912
U.S.A.

Mathematics Subject Classifications (1980): 41xx

Library of Congress Cataloging-in-Publication Data
Lagerstrom, Paco A. (Paco Axel)
  Matched asymptotic expansions : ideas and techniques / P.A.
Lagerstrom.
    p.   cm.—(Applied mathematical sciences ; v. 76)
  Bibliography: p.
  Includes index.

  1. Differential equations—Numerical solutions.   2. Asymptotic
expansions.   3. Singular perturbations (Mathematics) . I. Title.
II. Series: Applied mathematical sciences (Springer-Verlag New York
Inc.) ; v. 76.
QA1.A647 vol. 76
[QA372]
510 s—dc19
[515.3′5]   88-19958

Camera-ready copy prepared by the author using T$_E$X.

Printed in the United States of America.

9 8 7 6 5 4 3 2 1

ISBN 978-1-4419-3086-6

To
Saul Kaplun (1924–1964)

PREFACE

Content and Aims of this Book

Earlier drafts of the manuscript of this book (James A. Boa was then coauthor) contained discussions of many methods and examples of singular perturbation problems. The ambitious plans of covering a large number of topics were later abandoned in favor of the present goal: a thorough discussion of selected ideas and techniques used in the method of matched asymptotic expansions.

Thus many problems and methods are not covered here: the method of averaging and the related method of multiple scales are mentioned mainly to give reasons why they are not discussed further. Examples which required too sophisticated and involved calculations, or advanced knowledge of a special field, are not treated; for instance, to the author's regret some very interesting applications to fluid mechanics had to be omitted for this reason. Artificial mathematical examples introduced to show some exotic or unexpected behavior are omitted, except when they are analytically simple and are needed to illustrate mathematical phenomena important for realistic problems. Problems of numerical analysis are not discussed. Attempts at rigourous proofs are not favored; the author believes that a rigourous presentation of results of research, with no clarification of the essential ideas, is not always beneficial for advances in science, not even in mathematics. For similar

reasons the author has not attempted formal generality and abstractness; the aim
is to try to bring out the essential ideas by discussing concrete examples. There is
one omission in this book, as in other books on singular perturbations, which the
author regrets and attributes to his ignorance of relevant essential ideas: while there
are estimates of the order of errors there are no numerical error estimates. There
are other restrictions on the content. The discussion of problems involving partial
differential equations is sketchy, and delay equations, integral equations, etc., are
not discussed. The choice of topics is also influenced by the author's long collaboration with students, ex-students and colleagues from Caltech. Thus there has been
no attempt at completeness in the choice of topics or at a survey of the relevant
literature.

The central theme of the book is thus basic ideas and resulting techniques in
the method of matched asymptotic expansions (MAE), more specifically the concept
of matching. The point of view here is based on the fundamental ideas enunciated
by Saul Kaplun in the mid-fifties. In particular, matching is possible only when
the relevant expansions have a domain of overlap and matching is thus in essence
intermediate matching; the author is not aware of any counter example which is not
spurious. Another point is that the various recipes for matching can be explained,
and possibly corrected or discovered, from the basic underlying principles. The
author is not interested in formulating very general recipes for matching; however,
it is important to understand how showing that the term $C \log(x/\epsilon)$ in the inner
expansion can be matched with a constant term of order unity in the outer expansion
played an essential role in Kaplun's resolution of the Stokes paradox.

By necessity there will be some duplication of content with books on related
topics, in particular with those by Van Dyke, Cole, and Kevorkian-Cole. There
are three by now classical model equations for fluid mechanics, discussed here in
Chapter I, Sections 1, 4 and 5; the last two were introduced by the author. These are

discussed in various other books, however; the presentation here is more complete and, especially, introduces different points of view.

The author regrets many omissions in this book and also apologizes for faults of commissions which he has overlooked. However, further polishing would defeat its purpose. The long sequence of earlier drafts had to converge finally to a printed book, and the issue had to be forced by freezing the design.

Acknowledgements

The scientific collaboration with many students and with faculty during my long stay at Caltech has been of crucial importance for the preparation in this book.

J. A. Boa played an essential role as a coauthor of the initial drafts. For over four decades I have profited from exchange of ideas with J. D. Cole, my first coworker; much of our work was originally motivated by experimental research by H. W. Liepmann. I also have profited from collaboration with M. D. Van Dyke and with J. Kevorkian for almost as many decades. I worked with S. Kaplun during his entire scientific life; in addition to his published papers I have used many ideas of his communicated in private discussions; I regret having had to omit his profound work on generalizations of Poiseuille and Couette flow. W. L. Burke, another Caltech Ph.D., provided many stimulating discussions over many years; I regret omitting his pioneering perturbation analysis of the two-body problem in general relativity. Others were W. S. Childress and D. MacGillivray. W. L. Kath provided valuable comments on the manuscript. Many Caltech graduate students and undergraduates helped me; their specific contributions are mostly mentioned in the text. I would like to single out (with apologies to those unfairly omitted) the undergraduates Richard Holmes and Hanif Mamdani. It has been a special pleasure to work with some very young undergraduates, Paresh Murthy and Charles Fu, during the very last stages of the book. At Caltech, and during my stay in Paris 1960–61, I had many fruitful discussions with Paul Germain and Pierre Alais and other French scientists.

Competent and conscientious typing is of course essential. I can only single out Vivian Davies who did much of the original typing and Donna Gabai who skillfully guided the transition from older methods to the use of modern word processors.

PACO LAGERSTROM
Pasadena, Spring 1988

CONTENTS

CHAPTER III.

CHAPTER I
Introduction: General Concepts of
Singular Perturbation Theory

1.1. Approximations

Exact[†] solutions of mathematical problems are often difficult or impossible to obtain and one therefore often tries to find approximate solutions. There are many methods for doing this, and we shall briefly describe some of the methods. For a given problem, a mixture of various methods may be used.

One class of approximate solutions is those obtained by numerical calculations. Another broad class of approximate solutions give an approximate answer in closed

---

[†] We shall not indulge in useless quibbling of what "exact" and what "closed" mean. The function $u = \sin x$ may be the exact solution, in closed form, of a certain problem. However, for most values of x the value of u is irrational. It is thus never known exactly; on the other hand there are easy and well-defined methods for determining any desired number of decimals. Similarly, the notion of "in closed form" has been steadily extended by the elevation of more and more functions to the status of "known" functions. This concept is also vague: the more properties of a function that are discovered, the better known it is. If the solution is reduced to quadrature we also consider it known. A solution, approximate or not, involving only known functions is called an "analytic solution," or a solution in closed form. Poincaré made the pertinent observation that there are many mathematical problems (especially arising in applied mathematics) which one does not solve in the classical sense. One solves them "more or less." One would hope to solve them more and more. Similarly, an answer may be said to be more or less in closed form and further research may make it more in closed form.

analytical form. Among those one may compare iteration methods with perturbation methods. Iteration procedures normally consist in repeated applications of a functional operator to a function and successive operations are supposed to come increasingly close to the exact solution. A classical example is Picard's method, described in standard texts on partial differential equations. In some iteration methods one may often start with a rather arbitrary function, and the iteration leads towards the correct answer; in his instructive discussion of simple boundary-layer equations Weyl (1942) starts with the function which is identically zero. In a typical perturbation method one starts with a solution to a simpler problem and hopes that the solution obtained is close to the solution of the original problem. The solution so obtained is called the first approximation and often leads to equations which give higher order approximations, presumed to be even closer to the exact solution. *The basic principle of such perturbation methods is that neighboring problems have neighboring solutions.* The terms used are vague, even the term "neighboring solutions." Solutions $u(x)$ and $v(x)$ may be compared point by point. One may sometimes be interested in the relative error. Sometimes one may want to know only special properties of a function, e.g., the frequency of an oscillation, or the derivative at specific points (giving, say, aerodynamic skin friction or heat transfer), etc. The determination of *neighboring* or *related equations* is one of the major problems in singular perturbations. It will be discussed in detail in Chapter II, especially Section 1.

A desirable property of any approximation is an error estimate. If one approximates $\sin x$ by x for small values of x then the error is $< \left| \dfrac{x^3}{6} \right|$. This estimate is a numerical estimate. Another form of estimate is an order estimate. The difference $|\sin x - x|$ is of the order x^3. This is clearly a weaker and less useful estimate. (We shall define "order" below.)

The classical example of solving related problems is the sequence of approxi-

mations to π similar to the one used by Archimedes and his precursors. One starts
with neighboring problems, namely the circumscribed and inscribed squares for a
circle and finds the ratio of their perimeters to suitably defined diameters. (This
involves solving the equation $x^2 = 2$, and approximate solutions to this problem
can be obtained by solving the neighboring equations $x^2 = 1.96$ and $x^2 = 2.25$; the
estimates for $\sqrt{2}$ can be refined by choosing perfect squares closer to 2.)[†] The ratio
of the circumference of a circle to its diameter must lie between the corresponding
numbers for the inscribed and circumscribed squares. From these one may continue
to problems which approximate the circles, for instance inscribed or circumscribed
octagons, and so on. The use of polygons with more sides produces better approximations. These calculations are probably the earliest example of what are now
called pincer methods (Weyl's 1942 paper referred to above gives a recent example). One obtains upper and lower bounds for the answer, and hence estimates on
the error, that is an upper bound for the difference between the real value of π and
its approximations. Archimedes (Dover edition of Collected Works, p. 93) found
that $3 + \frac{10}{71} < \pi < 3 + \frac{10}{70}$. (To four decimals the bounds are 3.1408 and 3.1429
respectively whereas $\pi \doteq 3.1416$.)

1.2. Some Basic Concepts in Asymptotics

In perturbation problems one considers functions $u(x_1,\ldots,x_n \ ; \ \epsilon_1,\ldots,\epsilon_m)$ depending on two sets of arguments, one set $x_1,\ldots,x_n$ called variables, and another
set $\epsilon_1,\ldots,\epsilon_m$ called parameters. *The distinction between variables and parameters
is not intrinsic to the function but depends instead on the context.* In a physical
problem the distinction between coordinates and parameters is generally obvious.

[†] We do not insist on historical accuracy. The Greeks had only vague ideas about
irrational numbers.

From a mathematical point of view, if the function is implicitly defined by differential equations with boundary and/or initial conditions, the equations contain derivatives with respect to the coordinates but not with respect to the parameters. (However, in coordinate-type expansions, see Lagerstrom and Cole (1955, p.830) and Chang (1961, p.816ff.), a coordinate may play the role of a parameter.) For simplicity consider a single variable x and parameter ϵ. An approximation $f(x,\epsilon)$ to $u(x,\epsilon)$ will be a function, usually the solution of a simpler problem, which is uniformly close to $u(x,\epsilon)$ for x in some closed region R (to be discussed later) as ϵ approaches some distinguished value, which may be taken to be zero. Unless otherwise stated, it will always be assumed that the approach is through real positive values.[†] We shall say that $f(x,\epsilon)$ is an approximation to $u(x,\epsilon)$ uniformly valid to order $\varsigma(\epsilon)$ if

$$\lim_{\epsilon \to 0} \frac{u(x,\epsilon) - f(x,\epsilon)}{\varsigma(\epsilon)} = 0 \quad \text{uniformly for } x \text{ in } R . \qquad (2.1)$$

The function $\varsigma(\epsilon)$ is called a *gauge function*.

In asymptotic analysis it is necessary to determine the relative order of functions. To express this the Bachmann-Landau notation[‡] is very convenient ($\leftrightarrow$ means

---

[†] A deeper mathematical investigation may require a consideration of the whole complex ϵ-plane, even if only real positive values ϵ make physical sense. Van Dyke gives examples of regular perturbation problems for which the function is analytic in ϵ at $\epsilon = 0$ but the radius of convergence of the power series is finite due to a singularity located at a value of the parameter which has no physical meaning (say, a negative value of ϵ when only positive values of ϵ make physical sense). The nature and the location of the singularity may be estimated. By a transformation in ϵ the singularity may be removed to infinity and the radius of convergence, using the transformed parameter; see Van Dyke (1975, Note 13, and 1977). Similar techniques may be applied to improve the convergence of power series in x, for instance the Euler transformation (Van Dyke, 1964 and 1975). However, in singular perturbation problems one often encounters functions which are not analytic in ϵ at $\epsilon = 0$, a typical example being $e^{-x/\epsilon}$. Furthermore, one often deals with divergent series.

[‡] Somewhat modified here, *cf.* Kaplun (1967, p. 64 ff.). The subscript "s" indicates "strictly of the order of." The word "strictly" and the subscript "s" will often be omitted.

"if and only if").

$$\delta(\epsilon) = o\big(\eta(\epsilon)\big) \Leftrightarrow |\frac{\delta}{\eta}| \to 0 \text{ as } \epsilon \to 0 \, . \qquad (2.2a)$$

$$\delta(\epsilon) = O\big(\eta(\epsilon)\big) \Leftrightarrow \exists \text{ constants } C_1 \, , \, C_2 \text{ and } \epsilon_0 \, ,$$

$$\text{such that } \epsilon < \epsilon_0 \text{ implies } 0 < C_1 < |\frac{\delta}{\eta}| < C_2 \, , \qquad (2.2b)$$

$$\delta = O_s(\eta) \Leftrightarrow \lim |\frac{\delta}{\eta}| \text{ exists and is } \neq 0, \ \neq \infty \, . \qquad (2.2c)$$

For "$O_s(\eta)$" read "strictly of order η."

Since $\delta = O_s(\eta)$ is equivalent to $\eta = O_s(\delta)$ the symmetry of the concept is more clearly expressed by the notation

$$\delta \asymp \eta \Leftrightarrow \delta = O_s(\eta) \, . \qquad (2.2d)$$

If the limit in (2.2c) is equal to 1 we say that δ is asymptotic to η and write

$$\delta \sim \eta \, . \qquad (2.2e)$$

The functions $\delta(\epsilon)$ and $\eta(\epsilon)$ need only be defined in some interval $(0, \epsilon_0)$, ϵ_0 arbitrarily small but > 0. We shall restrict ourselves to functions of ϵ which are continuous, strictly positive and preferably monotonic in some interval $(0, \epsilon_0)$.

Sometimes it is convenient to use the notation of Hardy (1924), here somewhat modified,

$$\delta(\epsilon) \prec \eta(\epsilon) \Leftrightarrow \delta(\epsilon) = o\big(\eta(\epsilon)\big) \, , \qquad (2.3a)$$

$$\delta(\epsilon) \succ \eta(\epsilon) \Leftrightarrow \eta(\epsilon) = o\big(\delta(\epsilon)\big) \, , \qquad (2.3b)$$

$$\delta(\epsilon) \preceq \eta(\epsilon) \Leftrightarrow \delta(\epsilon) \prec \eta(\epsilon) \text{ or } \delta(\epsilon) \asymp \eta(\epsilon) \, , \qquad (2.3c)$$

$$\delta(\epsilon) \succeq \eta(\epsilon) \Leftrightarrow \delta(\epsilon) \succ \eta(\epsilon) \text{ or } \delta(\epsilon) \asymp \eta(\epsilon) \, . \qquad (2.3d)$$

The relation $\prec$ introduces an ordering among the functions considered. The ordering is only partial since $|\delta(\epsilon)/\eta(\epsilon)|$ need not tend to a limit. In that case

$\delta(\epsilon)$ and $\eta(\epsilon)$ are called *incommensurable*. However, *for practical purpose the use of commensurable functions, in particular the logarithmico-exponential functions of Hardy (1924), seems sufficient.* The relation $\asymp$ is transitive, symmetric and reflexive and hence an equivalence relation. The corresponding equivalence classes were denoted by ord δ by Kaplun (1967),

$$\text{ord } \delta(\epsilon) = \left\{ \eta(\epsilon) | \eta(\epsilon) \asymp \delta(\epsilon) \right\} . \tag{2.4}$$

These equivalence classes may be partially ordered in an obvious way. For this ordering symbols $<$, $>$, $\leq$, $=$ will be used.

The concept introduced by (2.2b) also leads to an equivalence concept, and a partial ordering of the corresponding equivalence classes

$$\text{Ord } \delta(\epsilon) = \left\{ \eta(\epsilon) | \eta(\epsilon) = O\big(\delta(\epsilon)\big) \right\} . \tag{2.5}$$

This is discussed in more detail in Kaplun (1967).[†]

An *asymptotic sequence* is a linearly ordered sequence of functions $\varsigma_j(\epsilon)$ such that

$$\varsigma_{j+1}(\epsilon) \prec \varsigma_j(\epsilon) , \quad \text{for each } j. \tag{2.6}$$

Typical examples are

$$1, \; \epsilon, \; \epsilon^2, \; \epsilon^3, \ldots, \tag{2.7a}$$

$$1, \; -\epsilon \, \log \epsilon, \; \epsilon^2 \log^2 \epsilon, \; -\epsilon^2 \log \epsilon, \; \epsilon^2, \ldots, \tag{2.7b}$$

$$\varphi(\epsilon), \; \varphi^2(\epsilon), \; \varphi^3(\epsilon), \ldots, \tag{2.7c}$$

with $\varphi(\epsilon) = |\log \epsilon|^{-1}$. We do not exclude functions which tend to infinity as ϵ tends to zero; a sequence may start with, say, $1/\epsilon$. Furthermore, we may have

---

[†] From the point of view of pure mathematics the study of the ord-spaces and Ord-spaces gives rise to many interesting theoretical problems. Some results are given by Kaplun (1967, p. 112ff.). This subject does not seem to have been pursued further and will not be discussed in detail in this book, an important exception being Kaplun's extension theorem (see Section 4). See also the final part of the present section.

multiple sequences, where the terms in the second sequence are *transcendentally small* (see below) with respect to the first. For instance, since

$$e^{-1/\epsilon} \prec \epsilon^n \ , \quad \text{all } n \ , \tag{2.8}$$

the sequence $(2.7a)$ might be followed by a sequence of powers of $(e^{-1/\epsilon})$, and $(2.7c)$ by a sequence of powers of ϵ.

Let $\varsigma_j(\epsilon)$, $j = 0, 1, 2 \ldots$ be an asymptotic sequence. We say that the sequence $a_j(x, \epsilon)$ is an asymptotic sequence of approximations to $u(x, \epsilon)$ uniformly valid with ς_j as gauge sequence in the closed x-interval D, iff for all j

$$\lim \frac{|u(x, \epsilon) - a_j(x, \epsilon)|}{\varsigma_j(\epsilon)} = 0 \ , \tag{2.9}$$

uniformly for all x in D. Note that the endpoints of the interval may depend on ϵ, which necessitates a trivial modification of the standard definition used for a fixed interval. Intervals with moving *i.e.*, ϵ-dependent intervals, are essential in singular perturbation techniques. Gauge functions are not unique, not even their order classes.

A very common method[†] for constructing the $a_j(x, \epsilon)$ is to find a sequence $\{f_j(x, \epsilon)\}$ such that

$$a_j(x, \epsilon) = \sum_{k=0}^{j} f_k(x, \epsilon) \ . \tag{2.10}$$

If (2.9) and (2.10) hold, the formal sum

$$\sum_{k=0}^{\infty} f_k(x, \epsilon) \tag{2.11}$$

is a *generalized asymptotic expansion* of $u(x, \epsilon)$ and we write

$$u(x, \epsilon) \sim \sum_{k=0}^{\infty} f_k(x, \epsilon) \ . \tag{2.12}$$

[†] In Chapter II Section 5, we shall encounter an example for which the most transparent form of the $a_j(x, \epsilon)$ is obtained by not writing $a_j(x, \epsilon)$ as a sum.

Properties of generalized asymptotic expansions are given by Erdélyi and Wyman (1963). Nothing whatsoever is said about the convergence of the series (2.11); the notation (2.12) is merely a restatement of the asymptotic result (2.9) with $a_j(x, \epsilon)$ given by (2.10). Often the series does not converge. Even when it does, it need not converge to $u(x, \epsilon)$. Olver (1970) has given a nontrivial example of this. An *asymptotic expansion in the sense of Poincaré, or of Poincaré type*, will be an expansion of the form

$$u(x, \epsilon) \sim \sum_{k=0}^{\infty} \beta_k(\epsilon)\, f_k(x)\,, \qquad\qquad (2.13)$$

where $\{\beta_k(\epsilon)\}$ forms an asymptotic sequence. The functions $\beta_k(\epsilon)$ will be called *expansion parameters*. They are usually found in the course of a solution of a problem but are not uniquely determined for a given problem. Expansions of Poincaré type have various nice properties not shared by generalized asymptotic expansions: they may be added, multiplied, integrated term by term, inverted, and usually differentiated (Olver 1974). Nevertheless, as we hope to show, their form is not sufficiently general for all the problems that can be handled successfully by perturbation techniques. We have just discussed three stages of generality. First we defined a sequence of functions $a_j(x, \epsilon)$ asymptotic in ϵ. Next we defined how each a_j may be defined as a partial sum of an expansion (2.10). Thirdly we specialized to Poincaré-type expansions (2.13). Occasionally, the more specialized form may be convenient for computations but it may then be put in a more general form which may be more desirable for various reasons.

For a Poincaré expansion one may define the concepts of *limit process* and *repeated application of a limit process*. The use of these concepts is familiar from the special case of a Taylor series. Let

$$u(z) = \sum_{j=0}^{\infty} u_j z^j\,. \qquad\qquad (2.14a)$$

Then the coefficients u_j are ("lim" denotes "limit as z tends to zero")

$$u_0 = \lim u(z) , \quad u_k = \lim z^{-k} \left(u - \sum_{j=0}^{k-1} u_j z^j \right) . \qquad (2.14b,c)$$

Thus u_k, which happens to be $\frac{1}{k!} \left(\frac{d^k u}{dz^k} \right)_{z=0}$, may be obtained by a repeated application of a limit process, using the z^k as expansion parameters. Similarly in the expansion (2.13) one may define, with suitably chosen β_k,

$$f_0(x) = \lim \beta_0^{-1} u(x,\epsilon) , \qquad (2.15a)$$

$$f_k(x) = \lim \beta_k^{-1} \left[u - \sum_{j=0}^{k-1} \beta_j(\epsilon) f_j(x) \right] . \qquad (2.15b)$$

Here limits are taken as ϵ tends to zero through positive values and the expansion parameters are considered to be known.[†] If we have determined $f_0 \ldots f_n$, we have the partial series $\sum_{j=0}^{n} \beta_j(\epsilon) f_j(x)$. We say that this partial series is obtainable by a repeated application of a limit process.

For any asymptotic sequence $\{\beta_k(\epsilon)\}$ there always exists a $\gamma(\epsilon)$ which is *TS* (*transcendentally small*) with respect to the sequence, *i.e.*,

$$\gamma(\epsilon) \prec \beta_k(\epsilon) , \quad \text{all } k . \qquad (2.16a)$$

This is proved by Hardy (1924) and in slightly generalized form by Eckhaus (1969). The standard example is

$$\gamma(\epsilon) \equiv e^{-1/\epsilon} = o(\epsilon^n) , \quad \text{all } n . \qquad (2.16b)$$

[†] The coordinate x is kept fixed in the limit processes. However, if $\tilde{x}$ is obtained from x by a scaling as explained in Section 3, one may define $u(x,\epsilon) = g(\tilde{x},\epsilon)$ and apply the limit process to g, keeping $\tilde{x}$ fixed and possibly using different expansion parameters. While, mathematically speaking, u and g are different functions, they represent the same physical quantity. Application of different limit processes to a physical quantity is an important tool in layer-type techniques, as will be explained in Chapter II.

One may then form a sequence of *transcendentally small terms (TST)*, say by positive powers of $\gamma(\epsilon)$. The process may of course be continued.

It is important to realize that the concept of transcendental smallness is an asymptotic one. Transcendentally small terms may be numerically important for finite values of ϵ, even if these values seem "small."[†] For example (2.16b) is not uniformly valid in n, in the sense that for a fixed $\epsilon < 1$, no matter how small, $e^{-1/\epsilon}$ is fixed while ϵ^n tends to zero with n. For numerical purposes it is instructive to compare $e^{-1/\epsilon}$ and ϵ^a for a fixed value of a and $0 < \epsilon < 1$. The smallest value of $\underline{a}$ for which equality occurs is easily found by assuming tangency of the corresponding curves. It is actually $e \doteq 2.718$.

$$\epsilon^e \text{ is tangent to } e^{-1/\epsilon} \text{ at } \epsilon = e^{-1} \doteq .3679 . \tag{2.17a}$$

Numerical calculations give the approximate bounds,

$$\epsilon^3 \leq e^{-1/\epsilon} \text{ for } .2205 \leq \epsilon \leq .5384 , \tag{2.17b}$$

$$\epsilon^4 \leq e^{-1/\epsilon} \text{ for } .1162 \leq \epsilon \leq .6994 . \tag{2.17c}$$

These results show that if TST are present but are ignored in calculations carried out to a moderate value of the exponent of ϵ, the results may be numerically unreliable.

We have little good advice to offer about estimating the numerical importance of terms transcendentally small with respect to a given asymptotic sequence. In the best of all possible worlds, each asymptotic approximation would be accompanied by a rigorous, *realistic* error bound. For linear ordinary differential equations of the second order, and certain of the methods for the asymptotic evaluation of integrals, such bounds may often be found. See, for instance, Olver (1974) and references

---

[†] It is a weakness of purely asymptotic theory that for numerical purposes one often cannot determine when a quantity is "small." Sometimes this may be partly remedied by numerical computing or by good numerical error estimates.

contained therein. However, even proofs of asymptotic validity of many of the expansions we construct for nonlinear and partial differential equations are lacking, although the expansions are undoubtedly correct. When proofs of validity do exist, the error estimates involved are often in the nature of order estimates; if a constant is involved, say $|u(x, \epsilon) - g(x, \epsilon)| < C\epsilon$, $C =$ known constant, the estimate is generally far from being the best possible. Sometimes the expansions we construct are rather good approximations to exact solutions constructed analytically or numerically even for moderate values of the formally small parameter. Some examples and counter examples will be given in Chapter II.

Ordering Properties of Real Numbers Compared with Those of Functions of ϵ. We shall here consider only the logarithmico-exponential functions of ϵ and let each order-class be represented by one of its functions. The real numbers will be positive; that is, we consider the interval $[0, \infty)$. Both the real numbers and the functions are linearly ordered. It is of both theoretical and, for our purposes, practical interest to compare the ordering. There is an isomorphism between the numbers in $[0, \infty)$ and the functions $\epsilon^{1/a}$, that is $a < b$ iff $\epsilon^{1/a} \prec \epsilon^{1/b}$ or between $(0, \infty)$ and the functions ϵ^a : $a < b$ iff $\epsilon^a \succ \epsilon^b$. However, the real numbers have ordering properties not shared by functions of ϵ. For instance, the numbers in the open interval $(0, 1)$ have a greatest lower bound $l = 0$ and a least upper bound $u = 1$. Consider all numbers $1/a$, > 0 , and let the sequence a_n converge to zero (for instance $a_n = 1/n$). Then for any number $b > 0$ there is an a_n such that $0 < a_n < b$, but there is no number b such that $0 < b < a_n$ for all n. Consider now the sequence ϵ^n, $n =$ positive integer. Obviously $n_2 < n_1 \Leftrightarrow \frac{1}{n_1} < \frac{1}{n_2} \Leftrightarrow \epsilon^{n_1} \prec \epsilon^{n_2}$. However, the decreasing sequence $\{\epsilon^n\}$ does not converge to the function which is identically 0. Between the sequence and the function which is identically zero one may insert any transcendentally small function. For instance $0 \prec e^{1/\epsilon} \prec \epsilon^n$ for all n. Another example is given by the functions $\epsilon^a |\log \epsilon|^b$ where $a =$ fixed number > 0, b any number > 0. If an increasing sequence a_n, $a_n < a$, converges to a then $\epsilon^a \prec \epsilon^a |\log \epsilon|^b \prec \epsilon^{a_n}$ for any $b > 0$, any a_n. A third example is $\varsigma_{(\epsilon)} = -1/\log \epsilon$ which has the property $\epsilon^a \prec \varsigma(\epsilon) \prec 1 = \epsilon^0$ for any $a > 0$. This example is related to the example of the transcendentally small terms since $\epsilon = e^{-1/\varsigma}$.

We shall not discuss the pure mathematical aspects of the above (which are interesting *per se*) but all of the examples given above are of practical importance and will actually occur in Chapter II.

1.3. Regular vs. Singular Perturbations.
Layer-Type vs. Secular Problems

When using a *regular perturbation technique* one often looks for a single Poincaré expansion, with expansion coefficients ϵ^j, of the function being approximated. The asymptotic sequence of expansion parameters is often $\{\epsilon^j\}$:

$$u(x, \epsilon) \sim \sum_{j=0}^{\infty} f_j(x)\epsilon^j . \qquad (3.1)$$

The function u may be defined by a differential equation $N(u, \epsilon)$ and, say, boundary conditions $B(u, \epsilon) = 0$; here B depends on the values of u and possibly some of its derivatives at the boundary points. One then substitutes the assumed form (3.1) of the expansion into the differential equation and boundary conditions, rearranges, and obtains the two series

$$\sum_{j=0}^{\infty} \epsilon^j L_j(f_0, \ldots, f_j) = 0 \text{ and } \sum_{j=0}^{\infty} \epsilon^j B_j(f_0, \ldots, f_j) = 0 . \qquad (3.2)$$

The B_j may involve derivatives of the f_k. The values of the f_k and its derivatives are evaluated at the boundary. It is then required that the coefficients of ϵ^j vanish separately. Thus the equations

$$L_j(f_0, \ldots, f_j) = 0 \text{ and } B_j(f_0, \ldots, f_j) = 0 , \quad j = 0, 1, 2, \ldots , \qquad (3.3a, b)$$

are obtained. For instance the second pair has the form

$$L_1(f_0, f_1) = 0 \quad \text{and} \quad B_1(f_0, f_1) = 0 , \qquad (3.3c)$$

in which f_0 appears as a known function. If (3.3c) can be solved, then f_0 is the first approximation and $f_0 + \epsilon f_1$ the second approximation to $u(x, \epsilon)$. Higher order approximations are determined recursively.

We may call singular[†] any problem for which the regular perturbation technique fails. In *layer-type* problems discussed in Chapters II and III, this happens because no obvious single asymptotic expansion of Poincaré type is uniformly valid throughout the x-domain of interest. In the *secular* (or *modulation*) problems, the non-uniformity occurs because of the cumulative effect of a small term acting for a long time. The following elementary[‡] but instructive example will illustrate how regular perturbation schemes may fail, and the difference between layer-type and secular problems.

Consider a damped linear mass-spring system with mass m, damping coefficient 2β, and spring constant k. It is governed by the differential equation

$$m \frac{d^2 u_d}{dt_d^2} + 2\beta \frac{du_d}{dt_d} + k u_d = 0 , \qquad (3.4)$$

[†] See, however, comment at the end of this section.

[‡] Cole (1968) introduced this example as an illustration. It uses physical ideas which are simple and known to anybody reading this book but which still illustrate many essential points, hence we have to describe it here without striving for originality. In particular the example can be used to show in great detail the use of dimensional analysis. In the rest of this book we usually assume that dimensional analysis has already been done, a notable example being the discussions of fluids (see Chapter II Section 6 and Chapter III Section 5).

Advice about scaling and nondimensionalization of equations has been given by Segel (1972) and Lin and Segel (1974). Although it is usually a good idea to use dimensionless variables to any given problem, sometimes it is helpful to retain dimensional forms. More precisely we shall show by examples (see *e.g.*, Chapter III Section 5) how one may use nondimensional reasoning while at the same time retaining the intuitive benefit of thinking in terms of dimensional quantities.

where u_d is the displacement from equilibrium and t_d the time elapsed, both in dimensional units. Initial values for the displacement and velocity at $t_d = 0$ will also be given.

We wish to write (3.4) in nondimensional form. To do so we must introduce dimensionless displacements and times by dividing the dimensional quantities by appropriate dimensional combinations of the parameters. If the initial amplitude is L, we may take as nondimensional displacement,

$$u = u_d/L \ . \tag{3.5}$$

(If the initial amplitude is zero, then a length can be defined from the initial velocity V and the parameters in (3.4) by $L = mV/\beta$). Actually, for a *linear* problem the scaling of the amplitude is not important. The parameters of the equation give two distinct parameters of dimension time. One is the reciprocal of the frequency of the undamped oscillator,

$$T_1 = (m/k)^{1/2} \ . \tag{3.6a}$$

As a second time we may take the damping time for a system with vanishing mass,

$$T_2 = \beta/k \ . \tag{3.6b}$$

T_1 and T_2 are functionally independent; by varying, say, β one can keep T_1 fixed and let T_2 vary. Other parameters of dimension time are, for instance,

$$T_1^a \, T_2^b \, , \quad a+b = 1 \ . \tag{3.6c}$$

Further examples will be given below. As a first dimensionless time variable we take

$$t = t_d/T_1 \ . \tag{3.7}$$

When u and t are used, (3.4) takes the nondimensional form

$$\frac{d^2u}{dt^2} + 2\epsilon \, \frac{du}{dt} + u = 0 \, , \tag{3.8a}$$

where

$$\epsilon = \frac{T_2}{T_1} = \frac{\beta}{(mk)^{1/2}} \; . \tag{3.8b}$$

Suppose that ϵ is very small, as will happen if the damping is weak compared to the mass and spring constant. An exact solution of equation $(3.8a)$ is

$$u = e^{-\epsilon t} \, \cos \sqrt{1 - \epsilon^2} \; t \; . \tag{3.9}$$

This solution contains two time scales of different order, t and τ,

$$\tau = \epsilon \, t = \frac{t_d}{T_3} \; , \quad \text{where } T_3 = T_1^2/T_2 = m/\beta \; . \tag{3.10}$$

These time scales are of different order, since $\tau = o(t)$ as $\epsilon \to 0$. We call t a fast time and τ a slow time, since when a constant t_0 is added to t, only ϵt_0 is added to τ. Obviously, variations on both scales are important for large values of t. We call this a *secular problem* (or *modulation problem*); obviously a simple expansion of the solution as a power series in ϵ would give unsatisfactory results. We shall only discuss such problems briefly; see Chapter II, Section 9.

Next suppose that ϵ is very large. Then any inverse power of ϵ is small. To find a suitable nondimensional parameter η and a nondimensional time variable s we shall use the trick of thinking dimensionally while at the same time using strictly nondimensional parameters: We see from $(3.8b)$ that ϵ^{-1} small may be thought of as either m small or k small, keeping the other dimensional parameters fixed. We first consider β and k fixed while letting m tend to zero. We thus want the small nondimensional parameter to multiply the first term while the coefficients of the second and third terms are independent of the parameters. In (3.4) we introduce the nondimensional u and divide by k to obtain

$$\frac{m}{k} \frac{d^2 u}{dt_d^2} + 2 \frac{\beta}{k} \frac{du}{dt_d} + u = 0 \; .$$

Thus we see that a suitable variable is

$$s = \frac{kt_d}{\beta} = \frac{t_d}{T_2} \; , \tag{3.11a}$$

and a suitable small parameter is

$$\eta = \frac{km}{\beta^2} = \frac{1}{\epsilon^2} \ , \tag{3.11b}$$

which gives the equation

$$\eta \, \frac{d^2u}{ds^2} \ + \ 2 \, \frac{du}{ds} \ + \ u = 0 \ . \tag{3.11c}$$

We now consider the second alternative. The spring constant k tends to zero while m and β remain fixed. The corresponding nondimensional parameter and variable may be found by applying the same reasoning as above, *mutatis mutandis*. Note that as the mass vanishes the first term in (3.4), or in (3.11c), drops out; the motion is entirely governed by the balance between damping and restoring force. However, suppose that we have an initial value problem with position and velocity prescribed at time zero. Both initial conditions cannot be satisfied for the first-order equation obtained by putting $\eta = 0$ in (3.11c). The only way the first term can be important as $\eta \downarrow 0$ is to have the second derivative increase to infinity; this also implies that order unity changes occur over an interval whose length tends to zero with η . Formally we may introduce

$$\tilde{s} = s\eta^{-1} \ . \tag{3.12a}$$

Introducing this variable into (3.11c) we find, after multiplying by η,

$$\frac{d^2u}{d\tilde{s}^2} + 2\frac{du}{d\tilde{s}} + \eta u = 0 \ . \tag{3.12b}$$

As $\eta \downarrow 0$ this gives an equation vanishing spring constant. Which equation, (3.11c) or (3.12b), should be used as $\epsilon \uparrow \infty$ and $\eta \downarrow 0$? If a singular perturbation technique is used, the answer is that both equations are needed. Clearly, (3.12b) has the advantage that two initial conditions may be imposed. The reasoning used above implies that (3.12b) is suitable initially. However, the acceleration, in terms of the variable s then is very large and it cannot stay large indefinitely. Thus we suspect that eventually the inertial term, proportional to m, becomes less important and (3.11c) should be used. The details of this technique will be discussed in detail in Chapter II. Here we shall make some preliminary remarks based on the exact general solution of (3.11c) which is

$$u = A_1 \, e^{\lambda_1 s} + A_2 \, e^{\lambda_2 s} \, , \tag{3.13a}$$

where

$$\lambda_1 = \frac{1}{\eta}(-1 + \sqrt{1-\eta}) \sim -\frac{1}{2} \, , \tag{3.13b}$$

$$\lambda_2 = \frac{1}{\eta}(-1 - \sqrt{1-\eta}) \sim -\frac{2}{\eta} \, . \tag{3.13c}$$

Again, two nondimensional time variables of different order, s and $\tilde{s} = s/\eta$, appear in the solution. However, for $s \geq 0$ the second term in (3.13a) is negligible (even transcendentally small) except in a narrow region, or layer, near $s = 0$. In this layer, which is roughly of width η, the term $e^{\lambda_2 s}$ drops rapidly from its value of 1 at $s = 0$ to nearly zero. More precisely, if $s \succ \eta$ then $e^{\lambda_2 s} \prec 1$; $0(1)$ changes in $e^{\lambda_2 s}$ occur on intervals, of nonnegative s , which are of size $0(\eta)$ measured in s-units. We shall say that $u(s, \eta)$ as given by (3.13a) has a *layer*, or layer of rapid transition of *width* $0(\eta)$ at $s = 0$. By this it is really meant that $u(s, \eta)$ changes rapidly in a thin layer, but the abbreviated nomenclature *layer* is convenient. When a layer occurs at a boundary point of the region considered, it will be called a *boundary layer*. If it occurs at an interior point, it will be called an *interior layer*. *Initial layer* would be an appropriate name for the layer in (3.13a). (The names boundary layer and shock layer had their origins in aerodynamics; see Chapters II and III.)

Layer-type vs. Secular Problems. Problems with solutions for which (3.13a) is representative will be called *layer-type* problems. They are similar to secular problems in that two or more scales of different order appear in the solutions. The major

difference is that the term $e^{\lambda_2 \tilde{s}}$, which uses the time $\tilde{s} = s/\eta$ may effectively

be neglected outside the initial layer. Thus the two scales do not need to be used

simultaneously, in contrast with the situation in (3.9). In both cases it may be

advantageous or necessary to use nondimensional time variables which are of the

same order as s and $\tilde{s} = s/\eta$, but have less simple forms. In (3.13a), $\hat{s} = -2\lambda_1 s$

and $s^* = -\frac{1}{2}\lambda_2 s$ would provide better results. Specific examples of such changes

of variable, which may loosely be called *coordinate transformations*, will be given

in the subsequent chapters. It is convenient to give here a definition of coordinate

transformation adequate for our purposes: Let x be a nondimensional coordinate

and ϵ the small nondimensional parameter. A coordinate transformation introduces

y and η by

$$y = \text{function of } x \text{ and } \epsilon \,, \quad \eta = \text{function of } \epsilon \,. \tag{3.14}$$

where $x = 0 \Rightarrow y = 0$ and $\eta \gtrless \epsilon$.

In order to see more clearly how a single Poincaré expansion will not ordinarily

produce a uniform approximation to a solution of a problem of layer type, consider

the function

$$u(x, \epsilon) = e^{-x/\epsilon} + x + \epsilon \,, \quad 0 \leq x \leq 1 \,. \tag{3.15}$$

This function is similar to (3.13a) but simpler. One application of the limit process

$\epsilon \to 0$, x fixed (called the *outer limit process*) produces an approximation

$$f(x) = x \,. \tag{3.16}$$

By using gauge functions $\varsigma(\epsilon) = 1$, we can see that $f(x)$ is a valid approximation

to order unity uniformly on intervals of the form $0 < x_0 \leq x \leq 1$, where x_0 is any

fixed positive constant. However, it is not uniformly valid on any interval containing

$x = 0$, because $u(x, \epsilon)$ has a boundary layer at $x = 0$. Let us rewrite (3.15) in terms of the *scaled* (or *stretched*) variable $\tilde{x} = x/\epsilon$. It becomes

$$u(\tilde{x}, \epsilon) = e^{-\tilde{x}} + \epsilon\tilde{x} + \epsilon , \quad 0 \le \tilde{x} \le 1/\epsilon . \tag{3.17}$$

Application of the limit process $\epsilon \to 0$, $\tilde{x}$ fixed (called the *inner limit process*) produces another approximation

$$g(\tilde{x}) = e^{-\tilde{x}} , \tag{3.18}$$

which is uniformly valid to order unity on intervals of the form $0 \le \tilde{x} \le \tilde{x}_0$, where $\tilde{x}_0$ is a fixed constant. However, it is not uniformly valid on the whole interval $0 \le \tilde{x} \le 1/\epsilon$, since if $\tilde{x}$ is of order $1/\epsilon$, then the term $\epsilon\tilde{x}$ has grown to order unity. Thus neither Poincaré approximation is uniformly valid on the whole interval. We shall see in the next section that the two approximations (3.15) and (3.17) together constitute an approximation which *is* uniformly valid.

Van Dyke (1964; 1975 p. 80 ff.) has stressed that if a problem contains two different dimensional times (or lengths or any other dimensional quantity), one much smaller than the other, then their ratio ϵ is a small parameter and perturbation techniques may be used. However, for the same reason one may form dimensionless times of different orders, and therefore singular perturbation techniques may be appropriate. This is not to be considered as an exact criterion but as a guideline. It is certainly not a necessary criterion. For example in the steady flow of a compressible fluid for which viscosity, heat conduction, and body forces are neglected, the important parameter is the Mach number M, which is the ratio of the speed of the fluid at infinity and the speed of sound at infinity. If M is small, a regular perturbation in M is used (Van Dyke 1975, pp. 15-16). However, if $M^2 - 1$ or $1/M$ are small, singular techniques may have to be used; see discussions of transonic flow and hypersonic flow in Cole (1968), Kevorkian-Cole (1981), Van Dyke (1964, 1975) and references given there.

In Chapter II, Section 9 we shall discuss briefly some other singular perturbation problems and techniques such as the Stokes-Lindstedt method, averaging, and the use of multiple scales.

Some layer-type expansions can also be converted into regular expansions by a coordinate transformation. An outstanding example is Kaplun's (1956) introduction of optimal co-ordinates, which will be discussed in Chapter III. Normally however, such changes are discovered only after a singular perturbation solution has already been determined. In Chapter II Section 5 we shall discuss an example for which a singular *or* a regular perturbation technique may be used, each method having distinct advantages.

Besides the possibility of converting singular problems into regular ones by changes of variables, there is also the possibility of solving problems by means not at all related to singular perturbation techniques. For example, we might be able to find exact solutions and their asymptotic expansions, or use iterative methods. Thus it is perhaps better to speak of singular perturbation techniques rather than problems. Furthermore, it is often possible to solve the same problem by several of the singular perturbation techniques mentioned above. Nevertheless we shall speak of singular perturbation "problems" when the techniques described in this book seem natural and convenient.

For some time it was thought that a necessary and sufficient condition for a problem to be a layer-type singular perturbation problem is that the small parameter multiply the highest order derivative. Historically, the pioneering investigations by Friedrichs and Wasow, originating in an analysis of Prandtl's solution of flow at high Reynolds numbers, mainly dealt with this case. However, the condition given, which is still occasionally encountered, is not only wrong but dangerously misleading; important counter-examples have been known for many decades. In the problem flow at low Reynolds numbers the highest derivative is not multiplied by the small parameter (see Chapter II Section 5). A realization that it is still a singular perturbation problem was essential for analyzing this problem and resolving its various paradoxes. On the other hand, there are examples[†] of equations having the highest derivative multiplied by a small parameter which, after an obvious transformation, can be solved by a regular perturbation method.

1.4. Domains of Validity. Overlap and Matching

Let us return to the function

$$u(x, \epsilon) = e^{-x/\epsilon} + x + \epsilon \,, \tag{4.1}$$

and two approximations to it,

$$f(x) = x \,, \tag{4.2a}$$

$$g(\tilde{x}) = e^{-\tilde{x}} \text{ where } \tilde{x} = x/\epsilon \,. \tag{4.2b}$$

These are valid to order unity on intervals of the form $0 < x_0 \leq x \leq 1$ and $0 \leq \tilde{x} \leq \tilde{x}_0$, respectively, where x_0 and $\tilde{x}_0$ are constants independent of ϵ. The second interval is $O(\epsilon)$ in width on the x-scale and hence shrinks to zero with ϵ. Closer inspection, however, shows that the estimates of validity are conservative. First, on the given intervals the expansions are actually valid to any order $\succ \epsilon$ (see (4.3) below). Second, it is possible to extend the interval on which $f(x)$ is valid to certain intervals whose left end point goes to zero with ϵ. This will be at the cost of having the error increase but the error will still be $\prec 1$ provided that the left end-point of the interval does not tend to zero too rapidly. For example, on the interval $\sqrt{\epsilon} \leq x \leq 1$, the maximum value of $u(x, \epsilon) - f(x)$ is $e^{-1/\sqrt{\epsilon}} + \epsilon$, which is

---

[†] See, for instance, the discussion of the Falkner-Skan equation with a large parameter in Lagerstrom (1964, p. 125 ff.).

still $O(\epsilon)$; thus $f(x)$ is still uniformly valid on that interval. However on the interval $\epsilon \le x \le 1$ the maximum error is $(1/e) + \epsilon$, and $f(x)$ is not uniformly valid to $O(1)$ on this interval. Similarly, $g(\tilde{x})$ is uniformly valid to $O(1)$ on $0 \le \tilde{x} \le \eta(\epsilon)$, where $\eta(\epsilon)$ is any function $\prec 1/\epsilon$. In terms of x, this says that $g(\tilde{x})$ is uniformly valid on any interval whose right endpoint shrinks to zero.

Consideration of intervals with moving endpoints is essential for understanding various singular perturbation techniques, in particular layer-type techniques. If $\mu(\epsilon) < \nu(\epsilon)$ one may define uniform convergence of $w(x, \epsilon)$ to zero on the closed interval $[\mu(\epsilon), \nu(\epsilon)]$ in the same way that one defines uniform convergence for a fixed interval $[a, b]$, a, b constants. One may also define uniform convergence in a domain D of order classes, defined by (2.4). Uniform convergence means that if M and N are order classes in D, with $M < N$, then $w(x, \epsilon)$ converges uniformly to zero in any interval $[\mu(\epsilon), \nu(\epsilon)]$ where $\mu(\epsilon)$ is an arbitrary function of M and $\nu(\epsilon)$ is an arbitrary function in N. One may also allow $M = N$ if one chooses $\mu(\epsilon) < \nu(\epsilon)$ for some $\epsilon < \epsilon_0 = $ constant. A *closed* order interval is of the form $D = \{Q | M \le Q \le N\}$. An open interval *may* have the form $D = \{Q | M < Q < N\}$. The definition of "open" involves concepts from topology which we shall avoid. Obviously, we would like the order domain $\{Q | M < Q < N\}$ to be open; we get a closed interval by adding the endpoints M and N. But consider the interval $D_{\alpha, \beta} = \{\mu(\epsilon) | \epsilon^{\beta} < \mu(\epsilon) < \epsilon^{\alpha}\}$ (we consider only logarithmico-exponential functions and let each function stand for its order class). Now let a and b be real numbers, $a < b$ and take the union D of all $D_{\alpha, \beta}$ such that $a < \alpha < \beta < b$. D should be considered to be open and may be called an interval, but it cannot be made into a closed interval by adding ϵ^b and ϵ^a. This follows from the discussion at the end of Section 2.

An approximation $f(x, \epsilon)$ is said to be an approximation uniformly valid in D to order $\varsigma(\epsilon)$ if

$$\frac{|u(x, \epsilon) - f(x, \epsilon)|}{\varsigma(\epsilon)} \to 0 , \quad \text{uniformly in } D . \tag{4.3}$$

If (4.3) holds, the corresponding statement is true if one replaces $\varsigma(\epsilon)$ by any $\varsigma^*(\epsilon)$, $\varsigma^*(\epsilon) \succeq \varsigma(\epsilon)$. Equivalently, $f(x, \epsilon)$ *is an approximation to* $u(x, \epsilon)$ *uniformly valid to order* $\varsigma(\epsilon)$ *in the domain of validity* D. The domain D may be contained in a larger domain of validity. The union of all such domains is the *maximal* domain of validity. The maximal domain of validity of $g(\tilde{x})$ as an approximation to $u(x, \epsilon)$ to order unity is

$$D_g = \{\eta(\epsilon)|\eta \prec 1\} \ . \tag{4.4}$$

D_f, the maximal domain of validity of $f(x)$ to order unity, is harder to characterize but certainly contains $\{\epsilon^a\}$ where $0 \leq a < 1$, and all functions o-equivalent to these. The intersection of a domain of validity of f with a domain of validity of g is called an *overlap domain to order unity*. In such an overlap domain both approximations are uniformly valid to order 1.

These definitions can be translated back into statements about x-domains of validity by picking a representative from D_f or D_g, say $\eta(\epsilon)x_\eta$ with x_η a fixed constant, and using it as the left or right endpoint of a moving interval. Clearly, picking one from each domain one can cover the entire x-interval $0 \leq x \leq 1$ by a suitable pair of intervals with moving endpoints. In this sense we can use the hybrid notation

$$D_f \ \cup \ D_g = [0, 1] \ . \tag{4.5}$$

The two functions $f(x)$ and $g(\tilde{x})$ together provide an approximation to $u(x, \epsilon)$ that is uniformly valid to order 1 on the entire interval $0 \leq x \leq 1$.

If two valid approximations to a function $u(x, \epsilon)$ have overlapping domains of validity, they are said to *match*. The technique of *matched asymptotic expansions* consists of constructing matching approximations which together give a uniformly valid approximation to some order throughout the x- domain of interest. Limit processes are an important tool in matched asymptotic expansions. The *outer* limit process applied to a function $u(x, \epsilon)$ is the standard one, in which ϵ tends to zero with x fixed. As we have seen, it is important to consider limits in which both ϵ and x tend to zero simultaneously. The *η-limit of* $u(x,\epsilon)$ is defined as the limit of $u(\eta x_\eta, \epsilon)$, as $\epsilon \downarrow 0$ while x_η, defined by $\eta(\epsilon)x_\eta = x$, is kept fixed. We shall denote this limit by $\lim_\eta$. Thus, the outer limit defined above is $\lim_1$ and the inner limit is $\lim_\epsilon$. When $u(x,\epsilon)$ is defined by (4.1), $\lim_\eta u = 0$ if $\epsilon^a = \eta \prec 1$ and $0 < a < 1$, $\lim_1 u = x$ and $\lim_\epsilon u = e^{-\tilde{x}} \equiv e^{-x/\epsilon}$. Some of the implications of η-limits have been explored by Meyer (1967) and Freund (1972).

Intermediate Matching. For constructing matched asymptotic expansions, the following easy lemma is of fundamental importance,

Lemma. *If $f(x,\epsilon)$ and $g(x,\epsilon)$ are approximations to $u(x,\epsilon)$ with an overlap domain D to order $\varsigma(\epsilon)$, and if η is in D then*

$$\lim_\eta \frac{|f(x,\epsilon) - g(x,\epsilon)|}{\varsigma(\epsilon)} = 0 . \qquad (4.6)$$

Formula (4.6) is used[†] in the following manner, called *intermediate matching*: In Figure 4.1 $f(x,\epsilon)$ may be valid to the right of $\eta_1(\epsilon)$ and $g(x,\epsilon)$ to the left of $\eta_2(\epsilon)$, so that $\eta(\epsilon)$ is in the overlap domain. The intermediate limit, $\lim_\eta$ is obtained by evaluating $|f(x,\epsilon) - g(x,\epsilon)|\varsigma^{-1}$ along the curve $x = \eta(\epsilon)$ as $\epsilon \downarrow 0$. Typically, in constructing expansions valid in certain domains we solve differential equations obtained by applying certain limit processes to the original full differential equation of the problem. The solutions may contain constants which are undetermined even after all appropriate boundary conditions have been applied. To determine the constants, we require that the two expansions match: that is, we search for a gauge function $\varsigma(\epsilon)$ and overlap domain D in which (4.6) holds. This can happen only if the constants take on certain values. Matching may also be needed to determine expansion parameters and, more generally, functions of ϵ occurring in an approximation.

[†] In principle. In practice the steps can be shortened considerably.

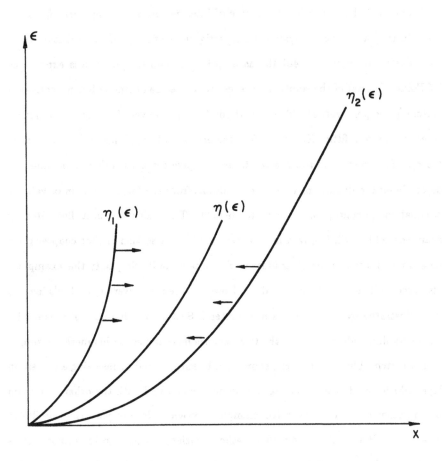

Figure 4.1. Intermediate Matching (Eq. 4.6).

While the principle of intermediate matching is clear it is difficult to determine *a priori* the domain of validity of an approximation. One valuable exact theorem exists, namely *Kaplun's extension theorem* (see below). This needs to be complemented by *Kaplun's Ansatz on the domain of validity*. However, often one simply resorts to brute force: Given two partial expansions, often called "outer" and "inner", one may check whether the recipe of intermediate matching works for some $\eta(\epsilon)$. In the early stage of systematic investigations of singular perturbation techniques (starting around 1940) the assumption of simple limit-process expansions (of Poincaré type) of the solution, together with the use of simple but unmotivated matching recipes, worked. These methods failed, however, for more complicated cases. In the mid-fifties Kaplun shifted the emphasis to applying *limit-processes to the equations rather than to the solutions* and gave certain heuristic guidelines for determining the domain of validity of solutions from the formal domain of validity of equations (a concept which is easily defined). These ideas are best illustrated by examples, which will be given in Chapter II. The first section of that chapter gives an example for which the older ideas work. Because of its simplicity this example is also used to illustrate the newer ideas. However, the full power of Kaplun's ideas is best illustrated by the examples of Chapter II Section 5. There are also examples where Kaplun's ideas fail, and the two expansions obtained by his methods simply do not match. This difficulty can sometimes be surmounted by inserting a transition layer which matches with one expansion on one end and with the other expansion on the other end. An illustrative example is provided by the relaxation oscillator at a van der Pol oscillator. Another method which sometimes works is to use terms in the equation which, following a stricter formalism, should not be retained. This often amounts to incorporating the solution for the transition layer into one of the expansions. The above examples will be discussed in Chapter II Section 8.

An Exact Theorem and an Ansatz. Most of the comments made in the latter part of the present chapter refer to the method of matched asymptotic expansions, best suited for layer-type problems. It seems fruitless to pursue an abstract discussion of methods any further here and we refer the reader to the concrete examples given in subsequent chapters. However, we shall now prove an extension theorem which definitely belongs to the present chapter.

Kaplun's Extension Theorem. *Let M and N be two order classes with $M \leq N$ and $f(x, \epsilon)$ an approximation to $u(x, \epsilon)$ valid to order $\varsigma(\epsilon)$ in the order domain $[M, N]$. Then there exist order classes $M_e < M$ and $N_e > N$ such that $f(x, \epsilon)$ is an approximation to $u(x, \epsilon)$ valid to order $\varsigma(\epsilon)$ in the extended order domain $[M_e, N_e]$.*

PROOF: We shall only prove the existence of M_e. The same reasoning will apply, with obvious changes, to N_e. Also it will be assumed that M is the class of function of order unity

$$M = \text{ord } 1 = \{\eta | \eta \asymp 1\} . \tag{a}$$

This assumption will make the proof more transparent. The same ideas could be applied to an arbitrary M. In fact, (a) implies no loss of generality since any case may be reduced to this case by rescaling x with an arbitrary element of M.

Essential for the proof is that while M, as defined by (a), does not contain any function which tends to zero with ϵ, it contains a sequence of constant functions $a_n > 0$ which tend to zero with n, for instance $a_n = 1/n$. We define

$$w(x, \epsilon) = \frac{|u(x, \epsilon) - f(x, \epsilon)|}{\varsigma(\epsilon)} . \tag{b}$$

Consider the intervals $[a_n, C]$ where C is an arbitrary constant $> a_1$. (Since we are only concerned with extending $[M, N]$ to the left it is not necessary to have an element of N as a right endpoint. The proof is also valid when $M = N$.) For each a_n there is by assumption an ϵ_n such that $w < a_n$ in the rectangle bounded by the horizontal sides $\epsilon = 0$ and $\epsilon = \epsilon_n$ and the vertical sides $x = a_n$ and $x = C$. (See Figure 4.2). One may assume $\epsilon_{n+1} < \epsilon_n$ and $\epsilon_n \downarrow 0$ with n.

Now define $\mu(\epsilon)$ by

$$\mu(\epsilon_{n+1}) = a_n , \tag{c}$$

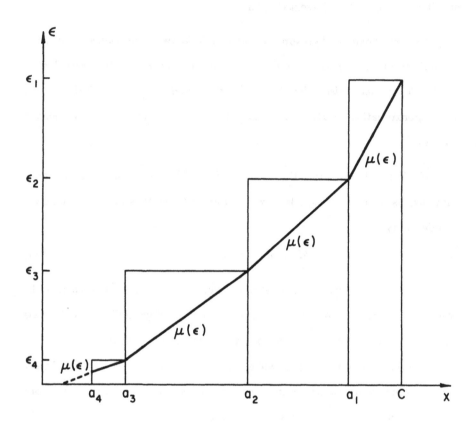

Figure 4.2. The Extension Theorem

and complete the definition by making $\mu(\epsilon)$ piecewise linear, as shown in Figure 4.2. Obviously $w(x,\epsilon)$ tends uniformly to zero in the interval $[\mu(\epsilon), C]$: Given an arbitrarily small $\delta > 0$, find an $a_n < \delta$. Then $w < a_n < \delta$ on any closed interval $[\mu(\epsilon), C]$ if $\epsilon \leq \epsilon_n$. Furthermore

$$\mu(\epsilon) \prec 1 . \tag{d}$$

Note that the rectangles on which $w < a_n$ have been replaced by regions bounded from above by $\epsilon = \epsilon_n$, from below by $\epsilon = 0$, from the right by $x = C$ and from the left by the curve $x = \mu(\epsilon)$.

It remains to be shown that w tends to zero uniformly in any interval $[\tilde{\mu}(\epsilon), C]$ if $\tilde{\mu} \asymp \mu$. It is sufficient to choose $\tilde{\mu}(\epsilon) = \mu_N(\epsilon) = \mu(\epsilon)N^{-1}$ for any positive integer N. To complete one replaces $\mu(\epsilon)$, for any fixed N, by

$$\mu_N(\epsilon) = \frac{a_{n-1}}{N} \, . \tag{e}$$

Thus the order class M_e of the theorem is

$$M_e = \text{ord } \mu(\epsilon) \, , \tag{f}$$

with $\mu(\epsilon)$ constructed as above. $\square$

Comments. Obviously one may continue and find an $M_e' < M_e$. In essence the extension theorem states that *a maximal domain of validity or of overlap is never closed.* Furthermore, the proof implies that if w tends to zero uniformly in $[\mu(\epsilon), \nu(\epsilon)]$, one may replace μ and ν by any $\tilde{\mu}$, $\tilde{\nu}$ of the same order.[†] We have already seen that $\varsigma(\epsilon)$ may be replaced by any $\tilde{\varsigma}(\epsilon)$ of the same order.

The extension theorem was first published in Kaplun-Lagerstrom (1957). The idea of the theorem, its proof and its application are, however, due to Kaplun. The proof given (*loc. cit.*) is concentrated but contains all the essential ideas. The theorem played an essential role in Kaplun's theory of flow at low Reynolds number, in particular his resolution of the Stokes paradox. Subsequently Lagerstrom (1961) constructed a comparatively simple model equation which can be used to illustrate Kaplun's ideas. It is discussed in detail in Chapter II Section 5.

Further discussion of the ideas of domain of validity, overlap, etc., will be given in Chapter II Section 1. See in particular the subsection on "The Method of

---

[†] There is of course a trivial exception to the last statement. If the function to be determined is defined for x in $[0,1]$ and one has an outer expansion valid, say, in $[\sqrt{\epsilon}, 1]$ one should not replace the right endpoint by any other constant or function. See Chapter II Section 1.

Kaplun" and following subsections. The discussion there is of a very general nature, but it is easier to understand when tied to a specific example. In particular we shall introduce an idea, necessary to complement the use of the Extension Theorem. We shall call this idea *Kaplun's Ansatz about the Domain of Validity*. It is given by (1.29) of Chapter II Section 1. We use the German word "Ansatz"[†] deliberately. It is an abomination to a purist: Not only is it not proved but it is unprovable; taken too literally it is false since counterexamples are easily found. Still it contains a genuine idea of great practical importance.

[†] "Heuristic idea" is an approximate although not very satisfactory English translation.

CHAPTER II
Layer-type Problems.
Ordinary Differential Equations

2.1. Friedrichs' Model Example and Simple Variants

This classical and historically important example was introduced by Friedrichs
(1942), primarily to illustrate that Prandtl's matching principle for boundary-layer
equations in fluid dynamics[†] made good sense in spite of its seemingly paradoxi-
cal nature. It has been extensively discussed and generalized in the literature. We
shall use the same form as Lagerstrom and Casten (1972), but the discussion will be
different. Also, since the understanding of singular perturbation problems has developed greatly during the past thirty-five years, our purpose will be different from
that of Friedrichs'. Rather than using the example to show that Prandtl was right
(incidentally, Prandtl's ideas about higher-order approximations were not correct),
we use it as an introductory problem whose solution is simple to obtain and has a
very simple structure. It is useful for illustrating various basic ideas and techniques.
In subsequent sections of this chapter we show how successively more sophisticated

---

[†] Knowledge of this subject is not necessary here. The equation has been called a model
equation for boundary-layer theory. However, it does not model the behavior of a fluid,
but is only concerned with the matching principle.

ideas must be used and how the structure of the solutions becomes increasingly more complex. Some of these ideas will be introduced here because they are easily grasped for this problem, even though not really necessary here. The significance and usefulness of these ideas will be apparent in subsequent problems.

The problem to be studied is

$$\epsilon \frac{d^2 u}{dx^2} + \frac{du}{dx} - a - 2bx = 0 \,, \tag{1.1a}$$

$$u(0) = 0 \,, \qquad u(1) = 1 \,. \tag{1.1b,c}$$

As always, except when explicitly noted, we assume $0 < \epsilon \prec 1$. The exact solution is

$$u(x, \epsilon) = (1 - a - b + 2\epsilon b)\frac{(1 - e^{-x/\epsilon})}{1 - e^{-1/\epsilon}} + ax + bx^2 - 2\epsilon bx$$

$$= \tilde{u}(x, \epsilon) + O(e^{-1/\epsilon}) \,, \tag{1.2}$$

where $\tilde{u}$ is obtained from u by neglecting the transcendentally small term $e^{-1/\epsilon}$ in the denominator of the first term.

Solution by matched asymptotic expansions. If we use a regular perturbation technique, *i.e.*, assume the solution to have the form $u(x, \epsilon) = \sum_{j=0} \epsilon^j u_j(x)$, we find that u_0 obeys a first-order differential equation; hence the two boundary conditions can be satisfied only accidentally. Thus, even to leading order, the term $\epsilon\frac{d^2 u}{dx^2}$ must play a role. This means that the second derivative and, as will be seen, also the first derivative, must be large in some region. If the linear homogeneous operator had been $L \equiv \left(\epsilon\frac{d^2}{dx^2} + 1\right)$ we would have rapid oscillations everywhere. But this cannot be the case for (1.1), since the roots of the characteristic equation for the homogeneous version of (1.1a) are both real. We therefore assume that there is a layer of rapid change somewhere. From now on we shall use the term "layer" for a "thin layer of rapid change." In the limit $\epsilon \downarrow 0$ this layer is expected to become a discontinuity. If the limiting point of discontinuity occurs at $x = x_d$ we may formalize the concept of rapid change by introducing a scaled variable $\tilde{x} = (x - x_d)\epsilon^{-s}$,

$s > 0$; ϵ^s is called the *scaling parameter* and the thickness of the layer is somewhat vaguely said to be of order ϵ^s. (In other examples it may be advantageous or even necessary to replace ϵ^s by a more general function $\eta(\epsilon)$.) In the layer x changes very little, and the essential behavior of the solution should be described by a function of $\tilde{x}$ whose derivatives are $O(1)$ on the $\tilde{x}$ scale. Now we have to determine s and x_d. We shall justify the choice

$$s = 1 , x_d = 0 \text{ and hence } \tilde{x} = x\epsilon^{-1}. \tag{1.3}$$

An essential condition is that in some sense the layer-solution (*inner solution*, whose leading term is denoted by $g_0(\tilde{x})$), can be joined to the solution in x obtained by neglecting the second derivative (outer solution whose leading term is denoted by $f_0(x)$). This joining is technically known as *matching*. To find an equation for g_0 we introduce $\tilde{x}$ into (1.1) and then let ϵ tend to zero where it appears explicitly, i.e., we keep $\tilde{x}$ fixed. This is known as applying the *inner limit* to the equation. It is desirable (and, as will be seen later, necessary) to have as full an equation as possible. This is achieved by putting $s = 1$. Thus, application of the inner limit to (1.1a) yields

$$\frac{d^2 g_0}{d\tilde{x}^2} + \frac{dg_0}{d\tilde{x}} = 0. \tag{1.4a}$$

The solution of (1.4a) is called the first term of the inner expansion (or inner solution). Its general solution grows exponentially as $x - x_d$ decreases and decays exponentially to a constant as $x - x_d$ increases.[†] As may be anticipated, and verified later, the first fact precludes the possibility of matching. This is the reason for putting $x_d = 0$. The solution of (1.4) should, vaguely speaking, be valid within a distance ϵ of $x = 0$. Thus it should satisfy (1.1b)

$$g_0(0) = 0 , \tag{1.4b}$$

[†] Later on (*e.g.*, Section 4) we shall encounter inner solutions which decay exponentially to constants both as $\tilde{x} \to -\infty$ and as $\tilde{x} \to +\infty$. The most frequently used are erf $\tilde{x}$ and tanh $\tilde{x}$.

but not necessarily (1.1c).

In the rest of the interval $\dfrac{d^2u}{dx^2}$ is not exceptionally large. Here then we get the equation

$$\frac{df_0}{dx} - a - 2bx = 0, \tag{1.5a}$$

with the boundary condition

$$f_0(1) = 1. \tag{1.5b}$$

Equation (1.5a) is obtained by taking the outer limit of (1.1a), i.e., letting ϵ tend to zero keeping x fixed. We call f_0 the leading term of the *outer expansion*. The variables $\tilde{x}$ and x are called the *inner* and *outer variables* respectively.

Since we now have only one boundary condition we find from $(1.5a, b)$

$$f_0(x) = 1 - a - b + ax + bx^2. \tag{1.6}$$

The original trouble with putting $\epsilon = 0$ in (1.1a) was that we obtained a first-order equation with two boundary conditions. This has now been resolved by giving up the inner condition $(u(0) = 0)$. However, for g_0 the opposite type of difficulty occurs. Equation (1.4a) is a second order equation but has only one boundary condition. The solution is, with an unknown constant C_0,

$$g_0(\tilde{x}) = C_0(1 - e^{-\tilde{x}}). \tag{1.7a}$$

We shall find that

$$C_0 = (1 - a - b). \tag{1.7b}$$

This constant is determined, not by applying another boundary condition, but by *matching the outer and inner expansions*. Prandtl introduced a recipe for matching which in the present case reads

$$f_0(0) = g_0(\infty). \tag{1.8}$$

This seems to be a flagrant contradiction with the previous reasoning. The outer solution f_0 is not supposed to be valid at $x = 0$. We have, however, extended it to zero by continuity, thus $f_0(0) \neq 0$. Even worse, g_0 was only supposed to be valid in a region of thickness ϵ, thus not at the right endpoint; furthermore $\tilde{x} = \infty$ lies outside the interval $[0,1]$. Without knowing Prandtl's reason for (1.8) one can get a vague explanation by saying that there is a region when x is very small but $\tilde{x}$ very large (for ϵ very small) and where both $f_0(x)$ and $g_0(\tilde{x})$ are valid. Boldly replacing "small" by "zero" and "large" by "infinity" one obtains (1.8). This reasoning contains the germ of Kaplun's theory of intermediate matching in an overlap domain. Formally, one may express (1.8) by stating that the outer limit (x fixed, $\epsilon \downarrow 0$) of the inner solution is equal to the inner limit ($\tilde{x}$ fixed, $\epsilon \downarrow 0$) of the outer solution.

It is desirable to have a solution which is valid uniformly everywhere to order unity. Simply adding f_0 and g_0 gives an error of order unity in the boundary condition. The reason is that f_0 and g_0 have a common part, namely $(1 - a - b)$. Subtracting this from the sum of f_0 and g_0 one gets the *composite* expansion to leading order

$$u_0 = C^{(0)}[u] = 1 - a - b + ax + bx^2 - (1 - a - b)e^{-\tilde{x}}. \qquad (1.9)$$

There is an alternative way of obtaining (1.9): The first *difference function* is defined by

$$v = D^{(0)}[u] = u - f_0. \qquad (1.10)$$

Then (1.1) may be written

$$\epsilon \frac{d^2v}{dx^2} + \frac{dv}{dx} + 2\epsilon b = 0 , \qquad (1.11a)$$

$$v(0) = -(1 - a - b), \quad v(1) = 0 . \qquad (1.11b,c)$$

The outer solution is then to leading order identically zero. To leading order the inner equation for v is the same as that for u. As boundary conditions we use

(1.11b) and match to zero. One finds the leading order inner approximation to v to be

$$\tilde{g}_0(\tilde{x}) = -(1 - a - b)e^{-\tilde{x}}. \tag{1.12}$$

Such a layer may be called a *correction layer*. There is now no duplication with the outer solution since the latter is identically zero. Thus (1.12) is also the first approximation to v. Adding f_0 to v one obtains the approximation to u given by (1.9).

To find better ("higher-order") approximations to u one assumes an *outer expansion*

$$E^{(k)}[u] = \sum_{j=0}^{k} \epsilon^j f_j(x) \ , \tag{1.13a}$$

and an *inner expansion*

$$H^{(k)}[u] = \sum_{j=0}^{k} \epsilon^j g_j(\tilde{x}) \ . \tag{1.13b}$$

The ϵ^j are called *expansion parameters* and are integral powers of the stretching (or scaling) parameter ϵ. Inserting (1.13a) into (1.1a) and using (1.1c) one finds a hierarchy of equations which may be solved without reference to the inner expansion. The solutions are

$$f_0(x) = 1 - a - b + ax + bx^2 \ , \tag{1.14a}$$

$$f_1(x) = 2b - 2bx \ , \tag{1.14b}$$

$$f_k(x) = 0, \quad k > 1 \ . \tag{1.14c}$$

Similarly inserting (1.13b) into (1.1) rewritten in $\tilde{x}$-variables one finds

$$g_0(\tilde{x}) = C_0(1 - e^{-\tilde{x}}) \ , C_0 = 1 - a - b \ , \tag{1.15a}$$

$$g_1(\tilde{x}) = C_1(1 - e^{-\tilde{x}}) + a\tilde{x} \ , C_1 = 2b \ , \tag{1.15b}$$

$$g_2(\tilde{x}) = C_2(1 - e^{-\tilde{x}}) + b\tilde{x}^2 - 2b\tilde{x} \ , C_2 = 0 \ , \tag{1.15c}$$

$$g_k(\tilde{x}) = 0 \ , k > 2 \ . \tag{1.15d}$$

The constants C_j have been found by matching. As a matching recipe one may use an obvious generalization of (1.8),

$$E^{(k)} H^{(k)} [u] = H^{(k)} E^{(k)} [u] . \qquad (1.16)$$

Here the left-hand side, for instance, is formed by applying $H^{(k)}$ to $u(x,\epsilon)$, as in (1.13b), rewriting the result as a power series in ϵ with coefficients depending on x and then neglecting terms $\prec \epsilon^k$. Examples are given below. In the present case there is no ambiguity in defining $E^{(k)}$ and $H^{(k)}$. We shall call this recipe the *formal matching method*. It was used very early by Friedrichs and others (see Lagerstrom (1957)). We call it "formal" because, while it may be proved in special cases, it was introduced as a recipe not supported by intuitive explanations. For further discussion and generalizations see Van Dyke (1975, p. 220ff.). Kaplun's method of intermediate matching, to be discussed later in this section, is based on intuitive ideas which explain (1.16) and the modifications of it which are sometimes necessary. It also works in some cases for which (1.16) or obvious generalizations thereof do not work (see especially Section 5).

Note that the expansions (1.13a, b) are simple *limit-process expansions*. In the outer limit, denoted by lim_1, one keeps x fixed as ϵ tends to zero; in the inner limit, lim_ϵ, $\tilde{x}$ is fixed as ϵ tends to zero. (Limits may be applied either to equations or to functions.) If the solution is known one may form $E^{(k)}$ and $H^{(k)}$ by repeated applications of the respective limits (and this is consistent with the method of construction used above):

$$f_0(x) = lim_1 u = E^{(0)}[u] , \quad f_j(x) = lim_1 \frac{u - E^{(j-1)}[u]}{\epsilon^j} , \quad j > 1 , \quad (1.17a)$$

$$g_0(x) = lim_\epsilon u = H^{(0)}[u] , \quad g_j(x) = lim_\epsilon \frac{u - H^{(j-1)}[u]}{\epsilon^j} , \quad j > 1 . \quad (1.17b)$$

This procedure resembles the construction of power series at $z = 0$ for a function which is analytic in a neighborhood of that point with z corresponding to ϵ. However, our limits are one-sided: ϵ tends to zero through positive values (or in the complex ϵ-plane in some wedge-shaped region to the right of the imaginary axis. In general, solutions of singular perturbation problems are not analytic in ϵ. For the present problem this is easily verified by inspection of the exact solution (1.2); for fixed x the solution has an essential singularity at $\epsilon = 0$.)

There is evidently much duplication between the inner and the outer expansions. For instance, two terms of $\epsilon^2 g_2$, namely $bx^2 - 2b\epsilon x$, occur in f_0 and ϵf_1 respectively. This explains why the composite expansion contains no ϵ^2 term. It is

$$\tilde{u}(x, \epsilon) = C^{(2)}[u] = E^{(2)}[u] + H^{(2)}[u] - E^{(2)} H^{(2)}[u] = u_0(x, \tilde{x}) + \epsilon u_1(x, \tilde{x})$$
$$= [(1 - a - b + ax + bx^2) + \epsilon(2b - 2bx)]$$
$$+ [-(1 - a - b)e^{-\tilde{x}} - \epsilon 2be^{-\tilde{x}}] . \tag{1.18}$$

The verification of this formula is left to the reader in Exercise 1.1.

Transcendentally Small Terms. The function $\tilde{u}(x, \epsilon)$ given by (1.18) represents in this simple case the entire composite expansion in powers of ϵ. By its construction it satisfies the boundary condition at $x = 0$. However, the boundary condition at $x = 1$ is satisfied only by neglecting TST (transcendentally small terms), that is terms of order $\preceq e^{-1/\epsilon}$. As shown in Chapter I, $e^{-1/\epsilon}$ may be actually numerically important compared to powers of ϵ for values of ϵ which one would consider "small." Thus for numerical purposes it may be important to consider these terms. A natural way of continuing the approximation is to consider the difference w defined by

$$w(x, \epsilon) = u(x, \epsilon) - \tilde{u}(x, \epsilon) , \tag{1.19}$$

find the equation for w, and solve it by a perturbation procedure in which we consider ϵ constant but

$$\delta = e^{-1/\epsilon} \tag{1.20}$$

as the small parameter. If the forcing function $a + 2bx$ were replaced by a more general function $h(x) = \sum_{j=0}^{\infty} \epsilon^j h_j(x)$ the composite expansion of u in ϵ would have had infinitely many terms if $h_j(x) \neq 0$ for a sequence of j tending to infinity. It would then be advisable to truncate the expansion in ϵ for some suitable value of j and to define $w(x, \epsilon)$ as u minus the truncated series and then proceed as above. The present case is sufficiently simple so that we may leave it to the reader to fill in the details (Exercise 1.2).

The Method of Kaplun. Limits of Equations. Principal Limits. Friedrichs' model equation has many simple properties enumerated at the end of this section. As a consequence it may be solved by very simple techniques. Those, however, are inadequate for solving more difficult problems, in particular the Stokes paradox of fluid mechanics (see Section 5 of the present chapter and Chapter III, Section 5). Furthermore, no intuitive justification has been given for the techniques used above. The Stokes paradox and related questions were solved by ideas introduced by Kaplun in the mid-fifties. We shall discuss and illustrate these techniques for the present example. They are not needed here, but it is nevertheless useful to explain them here because the example is so simple and the analytical work quite trivial so that the reader can concentrate on the ideas unhampered by any computational difficulties. The discussion will be rather pedantic to begin with. Later on it will be illustrated how one may simplify the technique considerably, once it is understood.

The first step is to find the proper relevant equations. We shall consider all possible equations obtained by limit processes (with certain provisos to be discussed below).[†] A variable x_η is defined by

$$x_\eta = \frac{x - x_d}{\eta}, \tag{1.21}$$

[†] For nonlinear equations one also has to consider scaling of the dependent variable; see Section 4.

where η is a function of ϵ, assumed positive and defined in some interval $0 < \eta < \eta_0$. Equation (1.1a) is then rewritten using this variable and a limiting equation is obtained as ϵ tends to zero, keeping x_η fixed. (Before doing so one should multiply the equation by a suitable function of η to make the largest terms of order unity.) This is called *taking the η-limit* of the equation. The constant x_d is the value of x at the point where the discontinuity will occur if one applies the outer limit ($\eta = 1$) to the solution and may be referred to as the position of the layer. It might possibly depend on ϵ but this complication will not occur here.

If x_η is introduced as the independent variable, (1.1a) reads

$$\frac{\epsilon}{\eta^2} \frac{d^2 u}{dx_\eta^2} + \frac{1}{\eta} \frac{du}{dx_\eta} - a - 2b(\eta x_\eta + x_d) = 0 \ . \tag{1.22}$$

Choosing η of various orders and taking the η-limits, one finds the following table,

$$\eta(\epsilon) \succ 1 \ : \ 2bx_\eta = 0 \ , \tag{1.23a}$$

$$\eta(\epsilon) = 1 \ : \ \frac{du}{dx_\eta} - a - 2b(x_\eta + x_d) = 0 \ , \tag{1.23b}$$

$$\epsilon \prec \eta(\epsilon) \prec 1 \ : \ \frac{du}{dx_\eta} = 0 \ , \tag{1.23c}$$

$$\eta(\epsilon) = \epsilon \ : \ \frac{d^2 u}{dx_\eta^2} + \frac{du}{dx_\eta} = 0 \ , \tag{1.23d}$$

$$\eta(\epsilon) \prec \epsilon \ : \ \frac{d^2 u}{dx_\eta^2} = 0 \ . \tag{1.23e}$$

Only functions $\eta(\epsilon)$ which are compatible with ϵ and with 1 are used. This is the proviso mentioned above. Consideration of other functions $\eta(\epsilon)$ would lead to unnecessary complications. The case $\eta(\epsilon) \succ 1$ leads to a nonsensical equation. In fact, the limit does not make sense since as ϵ tends to zero, x_η fixed, x will eventually become > 1, i.e., lies outside the given interval. In the future such limits will be disregarded *for finite intervals*. Obviously (1.23b) and (1.23d) are equations distinguished in two ways: They are obtained for specific order choices of $\eta(\epsilon)$ (not order intervals). Furthermore (1.23b) contains (is *richer than*)[†] (1.23c) in the sense that applying the η-limit, $\epsilon \prec \eta(\epsilon) \prec 1$, to (1.23b) yields (1.23c). Similarly, (1.23d) contains (1.23c) and (1.23e). Thus we need only the two distinguished equations (1.23b) and (1.23d), neither of which contains the other.

The general solution of (1.23d) is

$$u = Ae^{-(x - x_d)/\epsilon} + B \ . \tag{1.24}$$

For $x < x_d$ the absolute value of u grows exponentially. As was indicated earlier and will be verified later this would preclude matching. Thus

$$x_d = 0 \ . \tag{1.25}$$

For representatives of the order classes of 1 and ϵ we simply chose those functions themselves (there may be cases for which this is not the best choice). Furthermore, instead of the simple scaling with η one could have used a more general coordinate transformation and obtained an x_η^* which is $0_s(x_\eta)$. Finally, the dependent variable has not been scaled. In subsequent sections it will be shown that it may be convenient, or even necessary, to abandon the simplifications referred to above.

Thus, we have obtained the outer equations by applying the outer limit $(\eta = 1)$ to (1.1a).

$$\frac{df_0}{dx} - a - 2bx = 0 \ , \tag{1.26a}$$

[†] The concept of *richer than* (see Kaplun-Lagerstrom (1957)) is fundamental and will be used on many occasions in this book. A more elaborate explanation is the following: Equation 1.23b is actually the outer equation since x_d will be determined to be zero. Its solution is f_0 and hence by the Extension Theorem its domain of validity overlaps with the domain of validity of (1.23c). The solution of the latter equation is a constant C_0 which is determined by matching with f_0. By the same argument (1.23d) has a region of overlap with (1.23c). The unknown constant of the solution of (1.23d) is determined by matching with (1.23c) to be C_0 which already had been determined by matching (1.23c) with (1.23b). This procedure seems unnaturally elaborate but a similar argument will be essential in Section 5.

and the inner equation by applying the inner limit $(\eta = \epsilon)$,

$$\frac{d^2 g_0}{d\tilde{x}^2} + \frac{dg_0}{d\tilde{x}} = 0 , \quad \epsilon\tilde{x} = x . \tag{1.26b}$$

These are equations (1.5a) and (1.4a) respectively. We have written f_0 and g_0 instead of u to indicate that the equations are valid only for the leading term of the outer and inner expansion respectively. It is clear that anybody familiar with the idea of η-limits of an equation could have written down (1.26a, b) directly without going through the detailed process used above.

The equations contained in (1.23b) and (1.23d) respectively were discussed above. With an obvious definition of "formal validity" the result is:

The domain of *formal validity* of (1.26a) is the order domain $\epsilon \prec \eta \preceq 1$. The domain of formal validity of (1.26b) is the order domain $\eta \prec 1$. The formal domain of overlap of the equations is the intersection of those two domains, namely

$$\epsilon \prec \eta(\epsilon) \prec 1 . \tag{1.27}$$

Matching. Solving the inner and outer equations one obtains $f_0(x)$ as given by (1.6) and $g_0(x)$ as given by (1.7a) but with C_0 as yet undetermined. As stated earlier, C_0 can be determined by matching. Kaplun's explanation of matching is the following: Matching implies a comparison of two or several functions. However, such a comparison should in principle be done in an η-domain where both f_0 and g_0 are valid.[†] We know the formal domain of validity of the equations for f_0 and g_0. This is easily defined exactly. But what is the actual domain of validity? The function $\eta(\epsilon)$ is in the *actual domain of validity to order* ς iff

$$\lim{}_\eta (u - f_0)/\varsigma = 0 . \tag{1.28}$$

[†] For an apparent exception, see Section 5.

Here ς is called a *gauge function*. The concept of gauge functions will be discussed later; for the present we consider the case $\varsigma = 1$. To find the actual domain, Kaplun introduced a heuristic principle, namely:

Kaplun's Ansatz about Domains of Validity. *An equation with a given formal domain of validity D has a solution whose actual domain of validity corresponds to D.* $\hspace{4cm}$ (1.29)

One of the central problems in understanding singular perturbation techniques is to achieve an appropriate interpretation of the phrase, "corresponds to" in (1.29). Kaplun assumed it to mean "is equal to." This, as we shall see, is too bold an assumption.

The intersection of the domains of (actual) validity of f_0 and g_0 is called the *overlap domain*. In the present case the *formal* overlap domain is $\{\eta | \epsilon \prec \eta \prec 1\}$. According to the heuristic principle, both f_0 and g_0 should be valid approximations to the solution of problem (1.1) in this domain, with an error that is $o(1)$. Thus the intermediate limits (η-limits with $\epsilon \prec \eta \prec 1$) of the difference of these approximations should be zero. Written in terms of the intermediate variable $x_\eta = x/\eta$, this difference is

$$f_0 - g_0 = 1 - a - b + a\eta x_\eta + b\eta^2 x_\eta^2 - C_0(1 - e^{-\eta x_\eta/\epsilon}) . \qquad (1.30)$$

In principle this difference goes to zero for three reasons. 1) Some terms in f_0 and g_0 cancel identically. These terms represent a duplication of the leading terms in the outer and inner expansions and must not be duplicated in the first term of the composite expansion. 2) Some terms vanish because they are small; that is they vanish as $\epsilon \downarrow 0$. 3) Finally, terms cancel identically only if an as yet undetermined constant (or constants) is given an appropriate value. *Determining the value of this constant is the essence of matching*.

In the present case no terms cancel identically before matching, although such terms will appear in higher order matching. However, (1.30) contains terms which are small in the overlap domain: Under the intermediate limit the terms $a\eta x_\eta$ and $b\eta^2 x_\eta^2$ go to zero. The term $\exp(-\eta x_\eta/\epsilon)$ goes to zero for all η in $\epsilon \prec \eta \prec 1$. For this range of intermediate limits, the exponential term is transcendentally small. Finally we get to the essence of matching: Only the constant $(1 - a - b) - C_0$ remains. It must be identically zero which determines C_0 to be $1 - a - b$. We see that the actual domain of overlap of f_0 and g_0 coincides with the formal overlap domain; however in general this is not true.[†] In practice one is satisfied if the expansions to be matched overlap to *some* order for *some* range of intermediate limits. Later on it will be emphasized that it is the *idea* of intermediate matching which matters rather than the explicit use of the x_η as in (1.30). If one uses a smaller gauge function which is $\prec 1$, say $\varsigma_j = \epsilon^a$, $0 < a < 1$ the overlap domain will shrink since one must assume $\eta \prec \varsigma_j$.

Composite Expansions. Higher-Order Approximations. From f_0 and g_0, with C_0 determined by matching one finds the composite expansion u_0 (1.9). We may then define

$$\bar{u}_1 = u - u_0 \, , \qquad (1.31)$$

and find that

$$\epsilon \frac{d^2 \bar{u}_1}{dx^2} + \frac{d\bar{u}_1}{dx} + 2\epsilon b = 0 \, , \qquad (1.32a)$$

$$\bar{u}_1(0, \epsilon) = 0 \, , \quad \bar{u}_1(1, \epsilon) = (1 - a - b)e^{-1/\epsilon} \, . \qquad (1.32b, c)$$

The boundary conditions are now zero within TST (transcendentally small terms). The forcing function is $2\, \epsilon\, b$; hence it is natural to assume

$$\bar{u}_1 = \epsilon\, \bar{u}_1 \, . \qquad (1.33)$$

[†] Compare the discussion at the end of Chapter II, Section 2 and also Exercise 1.3 whose solution implies that Equations $(4.21c, d)$ of Lagerstrom and Casten (1972) need a minor correction.

Then $\bar{u}_1$ satisfies (1.32a) except that the forcing function is now 2b, and the $\bar{u}_1(1,\epsilon) = \frac{1}{\epsilon}\bar{\bar{u}}_1(1,\epsilon)$ which still makes it transcendentally small. One may now proceed as above and find inner and outer solutions and a composite term u_1. This yields the solution $u_0 + \epsilon\, u_1$ which is the composite expansion given by (1.18). Repeating the procedure, one sees that the terms $\epsilon^k u_k$, $k > 1$ are zero; hence the solution has been found if transcendentally small terms (TST) are neglected.

A second way of finding higher order terms is to form the difference function described by (1.10) and then to find its boundary layer, called the correction, or perturbation, or defect boundary layer (Lagerstrom (1961), Cole (1968), O'Malley (1974)). This may be generalized to higher order by finding $E^{(k)}[u]$, defining

$$D^{(k)}[u] = u - E^{(k)}[u]\,,\tag{1.34}$$

and then constructing the inner approximations to $D^{(k)}[u]$.

A third method is to find the inner and outer expansions and match at each stage.

The choice of methods is often, although not always, a matter of convenience or personal taste. It will be seen, however, that in certain problems the first two methods are difficult, for instance, due to the difficulty of composite expansions (see, e.g., Section 5).

In an easy problem such as the present one the third method is probably the clumsiest. We shall, however, use it here to explain higher-order matching:

Inserting $u - f_0$ into (1.1) we find that its value at $x = 1$ is zero and that the only nonhomogeneous term of the equation is of order ϵ. It comes from $\epsilon\dfrac{d^2}{dx^2}$ applied to f_0. We therefore assume that $u - f_0 = \epsilon\, f_1 + o(\epsilon)$ and find

$$\frac{df_1}{dx} - 2b = 0\,,\tag{1.35}$$

which has the solution given by (1.14b). A similar procedure gives $g_1(\tilde{x})$ of (1.15b) with C_1 undetermined. In matching one considers the difference

$$D = f_0 + \epsilon\, f_1 - g_0 - \epsilon\, g_1 = (1 - a - b + ax + bx^2) + \epsilon(2b - 2bx)$$

$$- (1 - a - b)(1 - e^{-x/\epsilon}) - \epsilon\, C_1(1 - e^{-x/\epsilon}) - ax \,. \qquad (1.36a)$$

We shall now do intermediate matching but want to demonstrate that introduction of symbols x_η for intermediate variables is not always necessary. We notice three aspects of matching. *First*, certain terms cancel identically so that one may write D as

$$D = bx^2 + \epsilon(2b - 2bx) + (1 - a - b)e^{-x/\epsilon} - \epsilon\, C_1(1 - e^{-x/\epsilon}) \,. \qquad (1.36b)$$

The *second* aspect of matching is that the *purpose* of matching is to determine C_1. We find that if we put $C_1 = 2b$ then the term $\epsilon\, C_1$ and $\epsilon 2b$ cancel identically even after division by ϵ which we use as principal gauge function. Note that since $2b - C_1$ is a constant it could never tend to zero in an intermediate limit unless it is identically zero. The result is now

$$\frac{D}{\epsilon} = \frac{bx^2}{\epsilon} - 2bx + \frac{(1 - a - b + \epsilon 2b)e^{-x/\epsilon}}{\epsilon} \,. \qquad (1.36c)$$

This expression has to tend to zero by itself. This is the *third* aspect of matching and necessitates that $x \prec \sqrt{\epsilon}$. In terms of intermediate variables, one could introduce x_η by $\eta\, x_\eta = x$ and demand $\eta \prec \sqrt{\epsilon}$. This is often an irrelevant formalism, not necessary for using the idea of intermediate matching. It is, however, recommended for difficult problems.[†]

We shall now give three comments on the matching just performed. 1) The domain of overlap, to order ϵ, is the η- domain such that $\eta \prec \sqrt{\epsilon}$ but η large enough

---

[†] Persistent use of the x_η-notation seems to have led to the unjustified opinion that intermediate matching is unwieldy. Theoretical and practical aspects of matching will be discussed in subsequent sections.

so that $e^{-x_\eta \eta/\epsilon}$ tends to zero. This is somewhat smaller than the formal domain of overlap of the corresponding equations (Exercise 1.3) but still good enough to determine C_1 which, we repeat, is the purpose of matching. 2) The use of any gauge function $\varsigma(\epsilon) \succ \epsilon$ could not have been used to determine C_1. Of course, we could have used some $\varsigma = 0_s(\epsilon)$ but this would, in the present example, only have led to redundant complications. 3) On the other hand, we could have used certain gauge functions $\varsigma \prec \epsilon$. The domain of overlap would then shrink but C_1 would still be determined. However, since we have not considered terms of order ϵ^2 we should demand that $\varsigma \succ \epsilon^2$. We call ϵ the *principal gauge function*.

If one continues the procedure for f_2 and g_2 one finds that $f_2 = 0$, $C_2 = 0$. After this there are no more terms in the inner and outer expansions.

Comparison of Formal and Intermediate Matching. First consider matching of $f_0(x)$ and $g_0(\tilde{x})$. Introducing an intermediate variable x_η one finds

$$\lim {}_\eta f_0(\eta x_\eta) = f_0(0), \quad \lim {}_\eta g_0(\eta x_\eta/\epsilon) = g_0(\infty) . \tag{1.37}$$

Thus requiring $\lim_\eta (f_0 - g_0) = 0$ is identical to using Prandtl's method (1.8). Of course, if for an intermediate η, $\lim_\eta f = f(0)$, (which simply means that ηx_η tends to zero with η then *a fortiori* $\lim_\eta f = f(0)$. The corresponding remark applies to $\lim_\eta g(\tilde{x})$. Hence (1.37) or (1.8) may be formulated

$$H^{(0)} E^{(0)}[u] = E^{(0)} H^{(0)}[u] . \tag{1.38}$$

Next consider matching of two terms. For convenience we repeat (1.14a, b) and (1.15a, b) which may be condensed into

$$E^{(1)}[u] = 1 - a - b + ax + bx^2 + \epsilon(2b - 2bx) , \tag{1.39a}$$

$$H^{(1)}[u] = (1 - a - b)(1 - e^{-\tilde{x}}) + \epsilon \left[C_1(1 - e^{-\tilde{x}}) + a\tilde{x} \right] . \tag{1.39b}$$

(It is assumed that C_0 in $g_0(x)$ has already been found by matching to be $(1-a-b)$.)
One finds

$$H^{(1)}E^{(1)}[u] = 1 - a - b + ax + 2\epsilon b, \qquad (1.40a)$$

$$E^{(1)}H^{(1)}[u] = (1 - a - b) + ax + \epsilon\, C_1 . \qquad (1.40b)$$

(Both formulas have been written using the variable x; one could equally well have
chosen $\tilde{x}$). Equating the two expressions gives the old result $C_1 = 2b$. In intermediate matching we applied some intermediate η-limit to $\epsilon^{-1}\big(E^{(1)}[u] - H^{(1)}[u]\big)$ (cf.
(1.36c)).

The third aspect of this matching was that certain terms vanish by themselves.
This is true for $e^{-\tilde{x}} = e^{-\eta x_\eta/\epsilon}$. Now this is, of course, a fortiori true if we take
a larger η, even outside the overlap domain, for instance $\eta = 1$. This accounts
for the absence of exponential terms in (1.40b). It also applies to $\frac{bx^2}{\epsilon} = b\frac{\eta^2}{\epsilon}x_\eta^2$ in
(1.39a). Here we saw that η could not be too large $(\eta \prec \sqrt{\epsilon})$. But again, this is true
a fortiori if we take $\eta = \epsilon$ even if ϵ is outside the overlap domain. This accounts
for the fact that bx^2 appears in (1.39a) but not in (1.40a). The second aspect of
intermediate matching was that the difference of certain terms, when divided by
ϵ, will never tend to zero unless they are identically zero. This forced us to put
$C_1 = 2b$ (the main purpose of matching). Above it corresponds to equating $2\epsilon b$ of
(1.40a) with ϵC_1 of (1.40b). The first aspect was that certain terms must vanish
identically in (1.36c). Above, this corresponds first of all to the equality of the
constant terms of order unity in (1.40a, b). But this had already been achieved in
matching of f_0 and g_0. Secondly, the same term ax occurs in (1.40a) and (1.40b);
the term ax occurs in f_0 and the identical term $\epsilon a\tilde{x}$ in ϵg_1. These must be identical,
for there is no longer any unknown constant to choose. It is part of the duplication
of terms in the inner and outer expansions, which after all are expansions of the
same function. Here again there is no difference between intermediate matching and

formal matching. Thus, if we equate (1.40a) and (1.40b), formal matching gives the same result as intermediate matching. The latter explains the former. The essence of the explanation is that if the η-limit of a term vanishes for η sufficiently small (large), then it vanishes a fortiori if one replaces η by a smaller (larger) function, even if these are outside the domain of overlap. The choices made in the formal matching are $\eta = \epsilon$ and $\eta = 1$ respectively.

Obviously, it is hard to say whether one method is faster than the other. We shall have to return to the subject of matching for more difficult problems in subsequent sections.

Simple Variants of Friedrichs' Model Equation. We shall now consider two-point boundary-value problems which are closely related to Friedrichs' problem. The equations discussed below will all have constant coefficients. However, as will be seen later the study of equations with constant coefficients can give very important qualitative hints for the study of equations with coefficients depending on x (Section 2) or on u (the nonlinear equation of Section 5).

First, consider the problem

$$\epsilon \frac{d^2 u}{dx^2} + c \frac{du}{dx} = a(x); \ u(x_0) = A, \ u(x_1) = B \ . \qquad (1.41a, b, c)$$

We assume $a(x) = \sum_{k=0} a_k x^k$ and that any power series occurring converges when needed. As usual ϵ is positive and, to fix the ideas, we assume $x_0 < x_1$. Let us first assume $c > 0$. The problem is then a trivial modification of (1.1). The boundary layer will occur at the left endpoint $x = x_0$. The use of a generalized correction-boundary layer as defined by (1.34), or possibly by the limit of (1.34) as $k \to \infty$, simplifies the calculations very much. The verification of this statement is left to the reader (Exercise 1.4). For $c > 0$ ("positive damping") the boundary layer occurs at the left endpoint, $(x = x_0)$ and for negative damping at $x = x_1$. (The case of $c = 0$ will be discussed later.) For instance, let $c = 1$. A solution of the homogeneous equation is $e^{-x/\epsilon}$. Multiplication by a suitable constant yields $e^{-(x-x_0)/\epsilon}$. A suitable inner variable is the $(x - x_0)\epsilon^{-1}$ so that the exponential function decreases to zero fast as x moves from the boundary into the interior (x_0, x_1). Similarly $c = -1$ yields (constant $\cdot e^{x/\epsilon}) = e^{-(x_1-x)/\epsilon}$ which is unity for $x = x_1$ and decreases exponentially as x moves into the interior.

Assume now that we add a term, du, $d = $ constant, to the left-hand side of (1.41a). The characteristic roots of the homogeneous equation are now

$$\lambda_1 = \frac{c}{2\epsilon}\left[-1 + \left(1 - 4\frac{d\epsilon}{c^2}\right)^{1/2}\right] \approx -\frac{d}{c}, \tag{1.42a}$$

$$\lambda_2 = \frac{c}{2\epsilon}\left[-1 - \left(1 - 4\frac{d\epsilon}{c^2}\right)^{1/2}\right] \approx -\frac{c}{\epsilon}. \tag{1.42b}$$

The assumption $c \neq 0$ is essential. It is the second root which determines the boundary layer. As before, we see that positive (negative) damping implies a boundary layer at the left (right) endpoint. If for instance $c = 1$ and $x_0 = 0$ a suitable inner variable is $x\epsilon^{-1}$. However, here we have used only the crudest approximation to λ_j.

Next, we consider the case $c = 0$,

$$\epsilon\, u'' + du = a(x)\,. \tag{1.43}$$

The subcase $d < 0$ gives an interesting variant on (1.1); see Exercise 1.5. However, if $d > 0$, say $d = 1$, the situation changes radically. The basic solutions to the homogeneous equations are $\sin x^*$ and $\cos x^*$, where $\epsilon^{1/2}x^* = x$. Obviously, we have no longer a simple variant of (1.1).

Finally we consider a problem involving a third-order equation for whose solution only minor modifications of previous methods are needed:

$$\epsilon\, \frac{d^3u}{dx^3} - \frac{du}{dx} = a\,, \tag{1.44a}$$

$$u(0) = A\,, \quad u'(0) = B\,, \quad u(1) = C\,, \tag{1.44b, c, d}$$

A, B, C, a are constants independent of ϵ. Special solutions of the homogeneous version of Equation (1.44a) are $e^{-x/\sqrt{\epsilon}}$ and $e^{-(1-x)/\sqrt{\epsilon}}$, so that boundary layers are possible at both $x = 0$ and $x = 1$. The scaling parameter is $\sqrt{\epsilon}$ rather than ϵ. Defining

$$\tilde{x} = \frac{x}{\sqrt{\epsilon}} , \quad \bar{x} = \frac{1-x}{\sqrt{\epsilon}} , \tag{1.45}$$

one finds that the solution of the boundary value problem is

$$u = \left[C - A + a - \sqrt{\epsilon}\,(a+B) \right] e^{-\bar{x}}$$
$$- \sqrt{\epsilon}\,(a+B)e^{-\tilde{x}} - ax + A + \sqrt{\epsilon}\,(a+B) + 0_s\big(e^{-1/\sqrt{\epsilon}}\big) . \tag{1.46}$$

Thus the expansion parameters and the principal gauge functions are integral powers of $\sqrt{\epsilon}$. One could easily obtain (1.46) by the methods used for (1.1), except for the following problem: Since boundary layers are possible at both $x = 0$ and $x = 1$, which boundary condition should be imposed on the reduced equation? The only intrinsic difference between $x = 0$ and $x = 1$ is that two boundary conditions are given at x=0, and only one at $x = 1$. However, suppose there were a boundary layer of order unity at $x = 0$: $g_0(\tilde{x}) = De^{-\tilde{x}}$, $D = 0_s(1)$. Then the x-derivative at $x = 0$ would be $0_s(D/\sqrt{\epsilon})$, whereas we have assumed it to be $B = 0_s(1)$. Thus a boundary layer of order unity is excluded at $x = 0$. On the other hand, a boundary layer of order unity is permitted at $x = 1$; hence one expects the nonuniform convergence to order unity of the outer solution to occur there. Thus the reduced equation should satisfy the lowest boundary condition at $x = 0$, namely $u(0) = A$; it is therefore $-ax + A$. To satisfy the boundary condition at $x = 1$ one must add a correction boundary layer $(C - A - a)e^{-\bar{x}}$. Similarly, a correction boundary layer $\sqrt{\epsilon}(a-B)e^{-\tilde{x}}$ must be added to satisfy $u'(0) = B$. This, however, alters the value of the solution at $x = 0$, and so the constant $\sqrt{\epsilon}(a-B)$ should be subtracted from the outer solution. Finally, a boundary layer correction $\sqrt{\epsilon}(a-B)e^{-\bar{x}}$ must be added at $x = 1$.

General theorems on boundary value problems of this type for higher order equations are given, for instance, by Wasow (1965, §37).

Summary of Some Simple Properties of Friedrichs' Model Equation. It may be instructive to summarize some features of the discussion in this section. In subsequent sections it will be seen that some of these features are absent in more difficult problems.

1) The scaling parameter is $\delta = \epsilon^{1/m}$ where m is the difference of degree between the full equation and the reduced equation. (This has only been shown for $m = 1$ and $m = 2$.)

2) Only two types of expansions, inner and outer, are needed. There may be two inner expansions but they are located at different points and their scaling parameter is the same. Now no transition layers are needed (*cf.* Section 9).

3) The expansions used are of the Poincaré type and can thus be obtained by repeated application of limit processes.

4) The expansion parameters are integral powers of δ.

5) One needs no special grouping or numbering of the terms of the expansion. The $(n + 1)$-th terms of the outer and inner expansions have the simple form $\epsilon^n f_n(x)$ and $\epsilon^n g_n(\tilde{x})$ respectively. Thus $E^{(n)}[u]$ and $H^{(n)}[u]$ have a clearly defined meaning.

6) The principal gauge functions used in matching $E^{(n)}[u]$ and $H^{(n)}[u]$ are ϵ^n.

7) The actual domain of validity of the solution of an approximating equation is almost as large as the formal domain of validity of the equation. The correction is due to terms of the form $e^{-x/\delta}$.

8) The proper stretching parameters and the principal equations may be obtained by a systematic application of various η- limits to the full equations.

9) The layers (of rapid change) occur at either boundary. There are no interior layers. In the outer limit the layer becomes a discontinuity of the function

itself.

10) As one moves from a boundary point towards the interior of the interval the boundary layer (if there is one) decays exponentially to a constant.

11) The outer expansion can be found independently of the inner expansions, except possibly for additive constants.

Exercises for §1

1.1. Verify that $C^{(k)}[u]$ of (1.18) is formed with the aid of operators $E^{(k)}$ and $H^{(k)}$ as indicated. Is there a difference if one uses $k = 1$ or $k = 2$? Also verify that the final formula satisfies $(1.1a, b)$ and $(1.1c)$ within transcendentally small terms.

1.2. Find some transcendentally small terms to improve the solution given by (1.18).

1.3. For Friedrichs' problem find the formal domain of overlap of $f_0 + \epsilon f_1$ and $g_0 + \epsilon g_1$ from the relevant equations. Determine also the actual domain of overlap to order ϵ.

1.4. Let

$$\epsilon\, u'' - u' = a(x); \quad u(0) = 0, \quad u(1) = 1,$$

where $a(x) = \sum_{k=0} a_k x^k$; assume that all power series occurring converge.

  a) Find the complete outer expansion $f = \sum_{k=0} \epsilon^k f_k(x)$ without using any term of the inner expansion.

  b) Give the equation and boundary conditions for the difference function $v = u - f$.

  c) Find complete outer and inner expansion of v.

  d) From the above results give a composite expansion of u, complete within transcendentally small terms. Indicate why the present method is much less unwieldy than that of finding the outer and inner expansions of u.

e) Specialize the above calculations to the case $a(x) = e^x$. Give the outer expansion and composite expansion of u in succinct form (sum power series occurring whenever possible).

1.5. Let

$$\epsilon\, u'' - u = \sum_{k=0} a_k x^k; \quad u(0) = 0, \quad u(1) = 3 .$$

Find first the outer expansion $f = \sum_k \epsilon^k f_k(x)$, define the difference function $v = u - f$, give the equation and the boundary conditions for v and find v using boundary layer techniques.

2.2. Linear Equations with Variable Coefficients

Introduction. In the preceding section we illustrated some basic ideas and techniques of the method of matched asymptotic expansions, using a very simple example (with a trivial exact solution). In the remainder of the present chapter we shall discuss ordinary differential equations of gradually increasing complexity with the double aim of introducing further ideas and techniques of matched asymptotic expansions and, eventually, of exhibiting problems for which this perturbation technique appears as a very natural way of obtaining useful asymptotic expansions. In the present section we introduce a seemingly very simple generalization of the previous section; we discuss linear equations with *variable* coefficients. We shall see that while some of the previous methods generalize in an obvious way, we encounter some new phenomena, even for the relatively simple examples considered.[†]

Following our general policy of studying only problems which are as simple as possible as long as they exhibit the phenomena we wish to study, we restrict

---

[†] However, it is with nonlinear problems that the full power of the technique and the necessity for using a great variety of ideas is demonstrated. Many such problems will be discussed later, starting with Section 4.

ourselves to the class of problems

$$\epsilon u'' + a(x)u' + b(x)u = 0 , \qquad (2.1a)$$

$$u(x_0) = A , \qquad u(x_1) = B . \qquad (2.1b,c)$$

Here $x_0 < x_1$, prime denotes d/dx, ϵ occurs only where explicitly exhibited, and the functions $a(x)$ and $b(x)$ are as smooth as required.

There is extensive literature on the use of matched asymptotic expansions for the above problem; see for instance Kevorkian and Cole (1978) for a detailed discussion of various cases, and Bender and Orszag (1978) for a varied collection of exercises. Our discussion of the subject will be quite short.

Before dealing with the method of matched asymptotic expansions we shall briefly mention some very powerful and popular methods for obtaining asymptotic solutions to (2.1). First, we use the standard method for removing the first derivative by introducing the new dependent variable $w(x)$ by putting

$$u(x) = k(x)w(x) , \qquad k(x) = \exp\left[\frac{-1}{2\epsilon} \int_0^x a(t)dt\right] ; \qquad (2.2a,b)$$

this transforms (2.1a) into

$$\epsilon^2 w'' + \left(-\frac{a^2}{4} + \epsilon b - \frac{1}{2}\epsilon a'\right) w = 0 . \qquad (2.3)$$

In this form the *Liouville-Green method* is applicable, a method covered in the references just mentioned, in the concise and very readable booklet by Erdélyi (1956) and in the more extensive treatment by Olver (1974). These books also discuss *turning-point problems*, that is, cases for which the coefficient of w in (2.3) changes sign in the interval considered. This case occurs frequently in eigenvalue problems for the Schrödinger equation in infinite domains. In books on quantum mechanics these problems are usually treated by the WKB[†] method, which is a

---

[†] For Wentzel, Kramer, and Brilluouin; mathematicians like to add a fourth letter "J" for H. Jeffreys.

development of the Liouville-Green method. This subject is extensively covered in the literature, and the methods will not be discussed here. We only point out that Cole (1968, p.106 ff) showed how a turning-point problem may be treated by matching asymptotic expansions.

Use of Matched Asymptotic Expansions. Comparison with the case of constant coefficients[†] gives some positive hints for finding asymptotic expansions, and also some hints which are negative in the sense that they indicate that the ideas used in Section 1 are not applicable, or at least not sufficient. We may think of $(2.1a)$ as the equation for a linear mass-spring system (oscillator) with ϵ = mass, $a(x)$ = damping coefficient, $b(x)$ = coefficient of restoring force, x = time.

A. If $a(x)$, $b(x) > 0$ in $[x_0, x_1]$ the oscillator is overdamped (for ϵ sufficiently small). If $b(x) < 0$, the term "overdamped" may not seem appropriate since the restoring force is repulsive. However, this is not important for a finite interval and we shall still call the motion overdamped whenever $a(x) > 0$. We may guess that if $a(x) > 0$ the behavior of $b(x)$ is less important; $b(x)$ may even change sign without affecting the solution qualitatively. Also we guess that there is an exponentially decaying boundary layer at $x = x_0$. Using the transformation $x \to -x$ one sees that the case $a(x) < 0$ gives similar results, *mutatis mutandis*.

B. Assume now that $a(x)$ vanishes at one endpoint and is $\neq 0$ in the rest of the interval. Without loss of generality we may assume $a(x_0) = 0$, $a(x) > 0$ in $(x_0, x_1]$ and take $x_0 = 0$, $x_1 = 1$. For analyzing this case the nature of the zero at $x = 0$ is crucial. We shall consider only algebraic zeros, $a(x) \sim x^k$ for x small. As we shall see the cases $0 < k < 1$, $k > 1$, and the limiting case $k = 1$ are significantly different.

[†] As will be seen in Section 4 this method yields much less for nonlinear equations.

C. Finally let $a(x)$ change sign in the interval. We shall assume a single simple zero which we may put at $x = 0$: $a(0) = 0$, $a'(0) \neq 0$, and shall now assume $x_0 < 0 < x_1$. There are two radically different subcases. If $a(x_0) < 0$, $a(x_1) > 0$ (*Case C1*) no boundary layer is possible. In special cases there is a possibility of an interior layer; this will be discussed below. On the other hand, if $a(x_0) > 0$ and $a(x_1) < 0$ (*Case C2*) boundary layers occur at both end points. These may be joined to the outer layer which is the solution in the interior of the interval. This layer obeys a first-order equation but no boundary conditions are given. The general solution then contains a constant C and for a continuous set of values of C both boundary layers can be matched to the outer layer. Thus C remains undetermined. Going to higher-order approximations will in general not resolve the indeterminacy.[†] However, as we shall see, the solution must satisfy a global condition (a term to be defined). We shall later on (Chapter III) give other examples of the use of global conditions to determine a constant which eludes the layer-type techniques. In our case a variational principle may be used as a global condition.

We shall now discuss in detail special examples of the various cases. In Cases A and B we put $x_0 = 0$, $x_1 = 1$.

Case A. $a(x) > 0$ *in* $[0,1]$. Guided by Section 1 we assume a boundary layer of thickness ϵ at $x = 0$, and an outer expansion valid away from $x = 0$,

$$u \simeq \sum_{j=0}^{\infty} \epsilon^j f_j(x) \ . \tag{2.4}$$

The equations for the f_j and the boundary conditions at $x = 1$ are

$$a(x)f_0' + b(x)f_0 = 0 \ , \quad f_0(1) = B \ , \tag{2.5a,b}$$

$$a(x)f_j' + b(x)f_j = -f_{j-1}'' \ , \quad f_j(1) = 0 \ , \quad j \geq 1 \ . \tag{2.5c,d}$$

[†] See, however, Bender and Orszag (1978, p.482, problem 9.32).

Solutions by quadrature are straightforward; we give the results only for $j = 0$ and $j = 1$,

$$f_0(x) = B \exp \int_x^1 \frac{b(t)}{a(t)} \, dt \, , \tag{2.6a}$$

$$f_1(x) = \left\{ \exp \left[\int_x^1 \frac{b(t)}{a(t)} dt \right] \right\} \left\{ \int_x^1 \left[\frac{f_0''(t)}{a(t)} \exp \int_t^1 -\frac{b(s)}{a(s)} ds \right] dt \right\} \, . \tag{2.6b}$$

As inner expansion we may take

$$g(\tilde{x}) \simeq \sum_{j=0}^{\infty} \epsilon^j g_j(\tilde{x}) \, , \quad \tilde{x} = x\epsilon^{-1} \, . \tag{2.7}$$

Rewriting (2.1a) in $\tilde{x}$ and expanding in ϵ we find

$$\frac{d^2 g_0}{d\tilde{x}^2} + a_0 \frac{dg_0}{d\tilde{x}} = 0 \, , \quad g_0(0) = A \, , \tag{2.8a, b}$$

$$\frac{d^2 g_1}{d\tilde{x}^2} + a_0 \frac{dg_1}{d\tilde{x}} = -a_1 \tilde{x} \frac{dg_0}{d\tilde{x}} - b_0 g_0 \, , \quad g_1(0) = 0 \, , \tag{2.8c, d}$$

where $a_0 = a(0)$, $a_1 = a'(0)$, $b_0 = b(0)$. Matching takes the place of the second boundary condition. Expanding $f_0(x)$ for x small gives

$$f_0(x) = B \exp \left\{ \int_0^1 \frac{b(t)}{a(t)} dt - \int_0^x \frac{b(t)}{a(t)} dt \right\}$$

$$= k \exp \left\{ -\int_0^x \frac{b(t)}{a(t)} dt \right\}$$

$$= k \left\{ 1 - \frac{b_0}{a_0} x + 0(x^2) \right\} \, , \tag{2.9a}$$

where

$$k = B \exp \int_0^1 \frac{b(t)}{a(t)} dt \, . \tag{2.9b}$$

To leading order the inner solution is

$$g_0(\tilde{x}) = A + C_0(1 - e^{-a_0 \tilde{x}}) \, , \tag{2.10a}$$

where the constant C_0 is determined by matching to be

$$C_0 = k - A \, . \tag{2.10b}$$

Thus a composite expansion valid to order unity is

$$u_0 = f_0(x) + (A - k)e^{-a_0 x/\epsilon} . \qquad (2.11)$$

It consists of the outer solution plus a boundary layer correction. Continuing the calculation, one finds

$$g_1(\tilde{x}) = -\frac{1}{2}a_1(A - k)\tilde{x}^2 e^{-a_0\tilde{x}} - \frac{1}{a_0}(A - k)(a_1 - b_0)\tilde{x}e^{-a_0\tilde{x}}$$

$$+ C_1(1 - e^{-a_0\tilde{x}}) - \frac{b_0 k}{a_0}\tilde{x} . \qquad (2.12)$$

Matching $f_0 + \epsilon f_1$ and $g_0 + \epsilon g_1$ gives

$$C_1 = f_1(0) ; \qquad (2.13)$$

the term $-b_0 k\tilde{x}/a_0$ in (2.12) cancels identically with a term in (2.9a). The remaining terms in each expansion are $o(\epsilon)$ in the overlap domain, and so go to zero by themselves. The composite expansion valid to order ϵ is now

$$u_0 + \epsilon u_1 = f_0(x) + \epsilon f_1(x) + (A-k)e^{-a_0\tilde{x}} + \epsilon\left[-\frac{1}{2}a_1(A-k)\tilde{x}^2 - \frac{1}{a_0}(A-k)\tilde{x} - f_1(0)\right]e^{-a_0\tilde{x}}.$$

$$(2.14)$$

As is easily checked, the first two terms of the outer or the inner expansion are obtained by applying the appropriate limits to (2.14). This shows that we have a correct composite expansion. However an expansion without these properties can also be obtained as follows. Above we replaced $a(\epsilon\tilde{x})$ by $a_0 = a(0)$ in the equation for g_0; the next term in the power series $a(\epsilon\tilde{x}) = a(0) + \epsilon\tilde{x}a'(0) + \ldots$ appears in the equation for g_1, etc. This amounts to a strict application of the inner limit. However, it is legitimate to retain higher-order terms in the power series for the equation for g_0. In fact, we may use the full function $a(\epsilon\tilde{x})$ instead of $a(0)$. A similar reasoning applies to g_1. Thus we may replace (2.8) by

$$\frac{d^2 g_0}{d\tilde{x}^2} + a(\epsilon\tilde{x})\frac{dg_0}{d\tilde{x}} = 0 , \quad g_0(0) = A , \qquad (2.15a,b)$$

$$\frac{d^2 g_1}{d\tilde{x}^2} + a(\epsilon \tilde{x})\frac{dg_1}{d\tilde{x}} = -b(\epsilon \tilde{x})g_0 , \quad g_1(0) = 0 , \qquad (2.16a, b)$$

and so on. Since the coefficients of the left-hand sides of (2.15) and (2.16) are variable, the functions g_j will depend on ϵ explicitly as well as through $\tilde{x} = x/\epsilon$, and so the inner expansion is no longer strictly of Poincaré form. However it is not necessary to be tied to Poincaré expansions, nor to limit process expansions generally. Various examples will illustrate this later.

The solution of (2.15) is

$$g_0(\tilde{x}) = A + D_0 \int_0^{\tilde{x}} \exp\Big\{-\frac{1}{\epsilon}\int_0^{\epsilon t} a(s)ds\Big\}dt . \qquad (2.17)$$

The integral in (2.17) is convergent (as seen by comparison with $\int_0^\infty e^{-mt}dt$ where $m > 0$ is the greatest lower bound of $a(x)$ on $0 \le x \le 1$) and hence may be written $\int_0^{\tilde{x}} = \int_0^\infty - \int_{\tilde{x}}^\infty$. Matching with f_0 then gives

$$A + D_0 \int_0^\infty \exp\Big\{-\frac{1}{\epsilon}\int_0^{\epsilon t} a(s)ds\Big\}dt = k , \qquad (2.18)$$

where k is given by (2.9b). The composite expansion valid to order unity is now

$$u_0(x) = f_0(x) + \frac{(A - k)\int_{x/\epsilon}^\infty \exp\{-\epsilon^{-1}\int_0^{\epsilon t} a(s)ds\} \, dt}{\int_0^\infty \exp\{-\epsilon^{-1}\int_0^{\epsilon t} a(s)ds\} \, dt} . \qquad (2.19)$$

It differs from the previous composite term (2.11) in the way the boundary layer corrections are handled; outside of the boundary layer, both corrections are transcendentally small. One may compare the expressions (2.11) and (2.19). If ϵ is small and $\tilde{x}$ moderate, the integrand in (2.19) can be expanded as

$$\exp\Big\{-\epsilon^{-1}\int_0^{\epsilon t} a(s)ds\Big\} = \exp\Big\{-a_0 t - \Big(\frac{\epsilon a_1 t^2}{2} + O(\epsilon^2)\Big)\Big\} = e^{-a_0 t}\Big(1 - \frac{\epsilon a_1 t^2}{2} + O(\epsilon^2)\Big) .$$

Thus

$$\frac{\int_0^{\tilde{x}} \exp\{-\epsilon^{-1}\int_0^{\epsilon t} a(s)ds\}dt}{\int_0^\infty \exp\{-\epsilon^{-1}\int_0^{\epsilon t} a(s)ds\}dt} = 1 - e^{-a_0 \tilde{x}} + O(\epsilon) , \qquad (2.20)$$

and the two expressions are again asymptotically equivalent to the order considered. The effect of keeping the full expression for $a(x)$ in (2.19) has been to sum certain terms that appear in expanded form when limit processes are used. In this sense something has been gained; what has been lost is the simplicity of constant coefficients. It seems plausible that (2.19) gives a numerically better value than (2.11) although this is difficult to decide *a priori*.

Note that in discussing Case A we have simply used the ideas of Section 1. The use of Equations (2.15)ff. represents a rather trivial modification. The computation of higher-order terms is straightforward in principle although obviously analytically more cumbersome than in Section 1.

Case B. $a(0) = 0$. We now study the case for which $a(x)$ has one zero only, located at one of the endpoints. Without loss of generality we may assume $x_0 = 0$, $x_1 = 1$, $a(0) = 0$. In $(0, 1]$ $a(x)$ then has to be either > 0, or < 0. We shall assume $a(x) > 0$ and consider only the case when $a(x)$ has an algebraic zero at $x = 0$, that is, $a(x) = x^k + o(x^k)$ (excluding for instance the case $a(x) = x^{1/2} |\log x|$), and take the simplest possible equation consistent with the assumptions made. Thus the problem is

$$\epsilon\, u'' + x^k u' + bu = 0\,, \quad b = \text{constant}, \quad k > 0\,, \tag{2.21a}$$

$$u(0) = A\,, \quad u(1) = B\,. \tag{2.21b}$$

Obviously, x is a suitable outer variable. However, replacing $a(x)$ by $a(0)$, which is $= 0$, does not give us an acceptable boundary-layer equation. To find a suitable inner variable we use the method introduced in Section 1 for determining which limit processes give the principal equations (*cf.* the discussion of Equation (1.23). This will give us a correct inner variable and justify the choice of the outer variable. We put $x_\eta = x\eta^{-1}$ with $\eta = \epsilon^l$ into (2.21). We have assumed that the layer occurs at $x = 0$ and that η is algebraic in ϵ. These assumptions will be tested by the success of the calculations below; the relation between l and k is yet to be determined. After multiplying by a suitable power of ϵ to clear the second derivative we get

$$\frac{d^2u}{dx_\eta^2} + \epsilon^{l(1+k)-1} \, x_\eta^k \frac{du}{dx_\eta} + \epsilon^{2l-1}bu = 0 \; . \tag{2.22}$$

If $0 < k < 1$ (we call this Case B1) we choose

$$l = \frac{1}{1+k} \; , \quad \eta = \epsilon^{1/(1+k)} \; . \tag{2.23}$$

The first two terms of (2.22) are then of order unity and

$$k < 1 \; , \quad \epsilon^{2l-1} = \epsilon^{(1-k)/(1+k)} = o(1) \; . \tag{2.24a}$$

For Case B2, $k > 1$, the choice (2.23) would make the third term dominate. Instead we keep the choice $l = 1/2$,

$$k > 1 \; , \quad l = \frac{1}{2} \; , \quad \eta = \epsilon^{1/2} \; , \quad l(1+k) - 1 > 1 \; . \tag{2.24b}$$

The first and third terms are now of order unity and the second term is $o(1)$. Finally, there is the limiting case, called Case B3,

$$k = 1 \; , \quad l = \frac{1}{1+k} = \frac{1}{2} \; , \quad \eta = \epsilon^{1/2} \; , \quad 2l - 1 = 0 \; . \tag{2.24c}$$

In this case all terms are of order unity. We now discuss the three subcases in some detail.

Case B1. $(0 < k < 1)$.[†] The outer expansion is again assumed to be of the form (2.4) and the equations (2.5), suitably specialized, are still valid. We shall carry out explicit calculations for the first two terms ($j = 0$ and $j = 1$). From the equation for f_0 we find,

$$f_0'' = f_0(bkx^{-1-k} + b^2x^{-2k}) \; .$$

[†] The case $k = 1/2$ was discussed in Cole (1968) and is also treated by Kevorkian-Cole (1981). They use the special choice of boundary-values, $A = 0$ and $B = e^2$; this choice is convenient but has otherwise no special significance. They also choose $b = -1$. Here we shall discuss the role of the signs of $a(x)$ and $b(x)$ and the effect of varying the value of k .

Putting $f_1(x) = f_0(x)h(x)$ gives,

$$h' = -bkx^{-1-2k} - b^2 x^{-3k} , \quad h(1) = 0 .$$

Thus

$$f_0(x) = B \exp \left[b(1 - x^{1-k})/(1 - k) \right] , \qquad (2.25a)$$

$$f_1(x) = f_0 \left[\frac{b}{2}(x^{-2k} - 1) - b^2(x^{1-3k} - 1)/(1 - 3k) \right] . \qquad (2.25b)$$

The computations of $f_j(x)$ for $j > 1$ are in principle straightforward. A minor complication may, however, occur: The solution may contain the indeterminate expression $(x^0 - 1)/0$. This is seen for f_1 if $k = 1/3$; in this case the second term in (2.25b) should be replaced by $-b^2 f_0 \log x$. Such logarithmic terms cannot occur when k is irrational. We expect that it is possible to derive a general formula for this case and then to handle the indeterminate expressions for rational k by letting k approach a rational number (see the discussion of logarithmic switchback in Section 5).

For Case B1 (2.22) becomes

$$\frac{d^2 u}{d\tilde{x}^2} + \tilde{x}^k \frac{du}{d\tilde{x}} + \varsigma(\epsilon)bu = 0 , \qquad (2.26a)$$

where

$$\tilde{x} = x\eta^{-1} = x\epsilon^{-1/1+k} , \quad \varsigma(\epsilon) = \epsilon^{(1-k)/1+k} , \qquad (2.26b, c)$$

This suggests an inner expansion of the form

$$u \simeq \sum_{j=0}^{\infty} \varsigma^j g_j(\tilde{x}) . \qquad (2.27)$$

Note that the expansion coefficients ς^j are integral powers of the scaling factor η or of ϵ only for exceptional values of j and k. The $g_j(\tilde{x})$ satisfy

$$\frac{d^2 g_0}{d\tilde{x}^2} + \tilde{x}^k \frac{dg_0}{d\tilde{x}} = 0 , \quad g_0(0) = A , \qquad (2.28a)$$

and, for $j > 0$,

$$\frac{d^2 g_j}{d\tilde{x}^2} + \tilde{x}^k \frac{dg_j}{d\tilde{x}} = -bg_{j-1} , \quad g_j(0) = 0 . \tag{2.28b}$$

The first equation is solved by

$$g_0(\tilde{x}) = C_0 G(\tilde{x}) + A , \quad G(\tilde{x}) = \int_0^{\tilde{x}} \exp\left(-t^{1+k}/1 + k\right) dt , \tag{2.29a,b}$$

where C_0 is determined by matching with f_0. To facilitate the matching we write the integral in (2.29b) as $\int_0^{\tilde{x}} = \int_0^{\infty} - \int_{\tilde{x}}^{\infty}$ and make a change of variable of integration, $s = t^{1+k}/1 + k$. Then $G(\tilde{x})$ can be expressed in terms of the gamma function and the incomplete gamma function,[†]

$$G(\tilde{x}) = (1 + k)^{-k/1+k} \left[\Gamma(1/1 + k) - \Gamma(1/1 + k, \, \tilde{x}^{1+k}/1 + k)\right] . \tag{2.29c}$$

As $\tilde{x}$ tends to infinity the incomplete gamma function decays exponentially to zero; hence matching reduces simply to

$$g_0(\infty) = C_0(1 + k)^{-k/1+k} \Gamma(1/1 + k) + A = f_0(0) = Be^{b/1-k} . \tag{2.30}$$

For $j > 0$ we find $g_j(\tilde{x})$ by the recursive formula,

$$g_j(\tilde{x}) = \int_0^{\tilde{x}} [\exp(-s^{1+k}/1+k)] \left\{\int_0^s \left[-bg_{j-1}(t) \exp(t^{1+k}/1 + k) dt\right]\right\} ds + C_j G(\tilde{x}),$$
$$\tag{2.31}$$

where $G(\tilde{x})$ is defined above. We note that $G(\tilde{x})$ is a solution of the homogeneous equation associated with (2.28b).

Matching. A complication in matching occurs because of the difference in expansion coefficients in the inner and outer expansions: If $k = \frac{1}{2}$, $\frac{1}{3}$, $\frac{1}{4}$ then $\varsigma(\epsilon)$ is $\epsilon^{1/3}$, $\epsilon^{1/2}$, $\epsilon^{3/5}$ respectively. The case $k = \frac{1}{2}$ is discussed in detail in Kevorkian-Cole (1981); for other cases see Exercises.

[†] The incomplete gamma function is defined in Erdélyi, *et al.* (Vol. II, 1953) as $\Gamma(k, x) = \int_x^{\infty} e^{-t} t^{k-1} dt$. It may also be written as $E_{1-k}(x)$ where $E_l(x)$ is the exponential integral defined in Section 5.

Case B2. $k > 1$. As mentioned in the discussion of (2.22) the inner variable is now determined by (2.24*b*) and the equation in inner variables is now

$$\frac{d^2 u}{d\tilde{x}^2} + \epsilon^{(k-1)/2} \tilde{x}^k \frac{du}{d\tilde{x}} + bu = 0 , \quad \tilde{x} = x\epsilon^{-1/2} . \tag{2.32}$$

For $k < 1$ the third term was smaller than the second term; this situation is reversed for $k > 1$. For $b(x) < 0$ we may apply boundary-layer technique as in B1, with obvious modifications. However, if $b(x) > 0$ the $g_0(\tilde{x})$ has oscillating behavior. On the other hand we see from (2.3), that is, the Liouville-Green form of the equation, that for ϵ sufficiently small the solutions of the equation are not oscillatory no matter what the signs of $b(x)$ or of $a(x)$ are, as long as $a \neq 0$ in the interior of the interval. Thus we conclude that the layer technique used here is not suitable; however, other asymptotic techniques such as the Liouville-Green method (not treated in this book) may be applied.

Case B3. $k = 1$. As pointed out above (see the discussion of (2.24)), in the scaling of x to obtain an inner variable all three terms of (2.21*a*) become equally important. If we put $\tilde{x} = x\epsilon^{-1/2}$ then (2.21*a*) becomes independent of ϵ; the resulting equation is discussed in texts on special functions defined by second-order linear differential equations, a subject which will not be discussed here. However if the interval is extended to $[x_0, x_1]$, $x_0 < 0 < x_1$, there is an interior zero at $x = 0$, a case discussed below.

Case C. $a(x)$ Changes Sign in the Interval. Turning-Point Problems. We assume that $a(x)$ changes sign only once and that $a(x)$ has a simple zero at $x = 0$; we let $x_0 = -1$ and $x_1 = 1$, so that the interval considered is $[-1, 1]$.

Subcase C1. If $a(-1) < 0$, $a(1) > 0$ (this occurs for instance when $a(x) \equiv x$), reasoning used above shows that no boundary layer is possible at either endpoint. There is, however, the possibility of an interior layer. To investigate this for specific instances assume that

$$a(x) \equiv x , \quad b(x) \equiv b = \text{ constant} . \tag{2.33a, b}$$

Assuming some kind of a layer at $x = 0$ we find that the leading term of the outer expansion obeys the reduced equation and is

$$f_0(x) = C_{\mp} x^{-b} . \tag{2.34}$$

Here C, the constant of integration, takes on the value C_- for $x \leq 0$ and C_+ for $x > 0$. If $b = -1$ we find

$$f_0(x) = -Ax \text{ for } x \leq 0 , \quad f_0(x) = Bx \text{ for } x \geq 0 . \tag{2.35}$$

Thus f_0 is continuous but its derivative is discontinuous at $x = 0$. This discontinuity has to be smoothed out by a *derivative layer* $g_0(\tilde{x})$ at $x = 0$. As inner variable we take $\tilde{x} = x\epsilon^{-1/2}$ and require that its derivatives match with those of $f_0(x)$. Details are left for the reader (see Exercise 2.1). However, this simple result depends on the special choice of b; the use of layer techniques is not obvious for a general value of b; see Exercise 2.2.

Subcase C2. If now $a(-1) > 0$, $a(1) < 0$, an exponentially decaying boundary layer is possible at each endpoint. However, each one obeys a second-order equation and only one boundary condition. The outer solution obeys a first-order equation but no boundary condition. Since two matchings are not enough to determine the three unknown coefficients, the difficulty with applying layer technique is that the problem appears underdetermined.[†] Actually, the missing condition may be supplied by a *global condition* which is independent of perturbation theory.

[†] See, however, Bender and Orszag (1978, p.482, problem 9.32).

To illustrate the nature of the problem we start by considering the simple function[†]

$$u = e^{\frac{1}{2\epsilon}(1-x^2)}, \qquad 0 < \epsilon \prec 1, \tag{2.36}$$

in the interval $[-1, 1]$. The behavior of this function is interesting: At $x = -1$ its value is $u(-1) = 1$; as x increases the function decays exponentially towards zero. Thus it practically drops dead (according to the conventional interpretation of fast exponential decay) and stays so for most of the interval. However, as x arrives within a distance of order ϵ to $x = 1$, $u(x)$ suddenly recovers and $u(1) = 1$. Now let us see how these results may be derived by applying perturbation methods to the boundary-value problem

$$\epsilon u'' - xu' - u = 0, \qquad u(-1) = 1, \qquad u(+1) = 1. \tag{2.37a,b,c}$$

The exact solution is of course (2.36), a fact to be ignored in deriving the perturbation solution. Since $-x$ is positive at $x = -1$ and negative at $x = 1$ we find boundary layers, g_l and g_r, at the left and right endpoints and an outer solution in between. To leading order they are

$$g_l = -A + (1 + A)e^{-\tilde{x}}, \qquad \tilde{x} = \frac{1 + x}{\epsilon}, \tag{2.38a}$$

$$g_r = A + (1 - A)e^{-x^*}, \qquad x^* = \frac{1 - x}{\epsilon}, \tag{2.38b}$$

$$f_0 = \frac{A}{x}. \tag{2.38c}$$

The corresponding composite solution is

$$h_0 = \frac{A}{x} + (1 + A)e^{-\tilde{x}} + (1 - A)e^{-x^*}. \tag{2.38d}$$

The solution satisfies both boundary conditions, and both boundary layers match with the outer solution; however, none of these conditions gives us a clue to the value of A.

[†] The author thanks R. E. O'Malley for showing him this example.

We note that the solution has a singularity at $x = 0$ except when $A = 0$; also, symmetry considerations indicate that the solution must be a function even in x, which occurs only for $A = 0$. Thus considerations outside of perturbation theory give the value $A = 0$, which of course also checks with the exact solution (2.36).

The situation seems to be the following: We need some method outside perturbation theory to determine the value of A. However, the methods given above seem highly special and of little use in a more complicated case. We shall therefore change the third term in (2.37a) from $-u$ to $+u$, thereby eliminating the singularity in f_0. We now consider the problem

$$\epsilon u'' - xu' + u = 0, \qquad u(-1) = U_- , \qquad u(1) = U_+ , \qquad (2.39a, b, c)$$

where U_- and U_+ are ϵ-independent constants. The solution is to leading order, in the notation of (2.38),

$$g_l = -A + (U_- + A)e^{-\tilde{x}} , \qquad g_r = A + (U_+ - A)e^{-x^*} , \qquad (2.40a, b)$$

$$f_0 = Ax , \qquad h_0 = Ax + (U_- + A)e^{-\tilde{x}} + (U_+ - A)e^{-x^*} . \qquad (2.40c, d)$$

Again perturbation techniques do not determine A; furthermore f_0 does not have a singularity for any value of A. Actually, even if $U_- \neq U_+$, one may find an equivalent symmetric problem and hence determine A by symmetry (see Exercise 2.3). Obviously, other equations may be found which spoil the symmetry more thoroughly. However, rather than trying to invent more complicated equations we shall stick to the relatively simple equation (2.39a) and simply ignore symmetry methods.[†]

The principle which we shall discuss is the variational principle introduced by Grasman and Matkowsky (1977). In our case we see that with the aid of an

---

[†] Kevorkian-Cole (1981, Section 2.3.9) discusses (2.39) with $U_- = 1$, $U_+ = 2$ using the variational principle to be discussed here.

integrating factor we may write (2.39a) as

$$\epsilon\left(e^{-\frac{x^2}{2\epsilon}} u'\right)' + e^{-\frac{x^2}{2\epsilon}} u = 0 \,. \tag{2.41}$$

If we define a Lagrangian by

$$L = e^{-\frac{x^2}{2\epsilon}} \left(\epsilon \frac{(u')^2}{2} - \frac{u^2}{2} \right) \,, \tag{2.42}$$

and require $u(x)$ to be chosen so that

$$J = \int_{-1}^{1} L(u', u, x) dx \tag{2.43}$$

has a stationary value for all functions satisfying the boundary values (2.39b, c), then $u(x)$ must obey the Euler-Lagrange variational equation which for (2.42) is exactly (2.41). We have already restricted the possible functions to those defined to leading order by (2.40d). Now h_0, and hence J, depend on a single unknown parameter A. If we insert h_0 into (2.43) we get a function of one variable, $J(A)$. The value which makes J stationary is thus found from

$$\frac{dJ}{dA} = 0 \,. \tag{2.44}$$

It is left to the reader to carry out the details of the computation of A (Exercise 2.4).

Like the symmetry considerations the variational principle is a global principle. We remark that it has been applied to nonhomogeneous second-order linear equations for a problem in probability in Matkowsky and Schuss (1977). However the method is obviously not by its nature restricted to linear problems.

Discussion. In Cases A and B the ideas used for equations with constant coefficients are still applicable. However, the solutions obviously present more analytic difficulties; alternative methods, such as the Liouville-Green method, may be used to great advantage. In Case C (turning-point problem) we encounter radically new phenomena. When there is a boundary layer at each endpoint, an unknown constant of the outer expansion cannot be determined by matching. The constant can often be found by applying a global principle.

In Ackerberg-O'Malley (1970) resonance, another strange phenomenon for turning-point problems, is pointed out. The resolution of such resonance problems is rather intricate; we refer to Kreiss (1981) and various references given there.

Exercises for §2

2.1. Assuming $a(x) \equiv x$ and $b = -1$, carry out the details of the matching of the outer expansion and the derivative layer for Subcase C1. Discuss composite expansion and higher-order approximations.

2.2. Discuss how the solution of the previous exercise changes when b is given various constant values. What happens if $b(x)$ is a simple function of x.

2.3. Show that if $u(x)$ satisfies (2.39a) then so does $v(x) = u - Bx$, $B = \text{constant}$. Choose B such that the problem is symmetric (even) in v. From the solution for v determine the value of H for the original problem (2.39).

2.4. Carry out the details of solving (2.44) to find the value of A.

2.3. Two Artificial Examples to Illustrate Techniques

In Section 2 Friedrichs' example was generalized to linear equations with variable coefficients. It would be natural to consider nonlinear equations next. This will be done in Section 4 and in subsequent sections, where many new phenomena will be discussed. However, in the present section we make a detour to consider two equations both of which are artificial in the sense that they were not suggested by physical problems, and intuition will play but a small role in the discussion. Even so, the examples are not chosen to exhibit the various types of exotic behavior possible in perturbation solutions: The first example illustrates the occurrence of logarithmic terms in expansion coefficients, a phenomenon which will be encountered frequently later on, and also shows how a systematic search for suitable expansion coefficients can be made; the second example seeks to clarify the idea of matching by overlap.

First Example.[†] The problem is

$$(\epsilon + x)\frac{d^2 u}{dx^2} + \frac{du}{dx} = 1 \ , \tag{3.1a}$$

$$u(0) = 0 \ , \quad u(1) = 2 \ . \tag{3.1b, c}$$

It resembles a simple version of Friedrichs' problem. However, the factor multiplying the second derivative in (3.1a) is $(\epsilon + x)$ rather than ϵ. There are two important direct consequences of this. First, the reduced equation (obtained by putting $\epsilon = 0$) still contains a second derivative. Second, the reduced equation is singular at the left endpoint in the sense that the function multiplying the second derivative vanishes there. It will be shown that these properties of the equation lead to features of the solution not occurring in the preceding sections; they will, however, occur in subsequent sections for examples motivated by physical problems.

Example (3.1) was introduced by Eckhaus (1973, pp. 102–105, 137–138). He gives the exact solution

$$u(x, \epsilon) = x - \frac{\log(1 + \frac{x}{\epsilon})}{\log \epsilon - \log(1 + \epsilon)} \ , \tag{3.2}$$

and finds, correctly, that although ϵ is still an appropriate scaling parameter the expansion parameters cannot be exclusively powers of ϵ. He therefore abandons singular perturbation techniques and suggests an iteration procedure. We shall, however, show how one may use singular perturbation methods in a systematic way. In the course of this it will be found necessary to introduce expansion parameters

---

[†] The discussion of this example utilizes unpublished notes by J. A. Boa.

of the form $\epsilon^j |\log \epsilon|^k$. Actually, as will be shown in subsequent sections such expansion parameters occur naturally in many physical problems; see for instance Section 5 and Section 6 below, Lagerstrom-Reinelt (1984) and Chapter 10.5 of Van Dyke (1964, 1977).

Equation (3.1) suggests a shift in the independent variable $x \to y = x+\epsilon$ which would make ϵ disappear from the equation and reappear in the boundary condition. This method will, however, not be used since it is more instructive to keep the original x. Introducing an inner variable $(x - x_d)/\epsilon^a$, $a > 0$, one finds that if $x_d = O_s(1)$ and $\neq 0$ then $x_d \epsilon^{-a}$ would dominate the factor of the second derivative and lead to an unacceptable inner equation. Putting $x_d = o(1)$ would lead either to an unacceptable equation or unnecessary complication. As mentioned above, one could have put $x_d = -\epsilon$ but we prefer to avoid this choice. Thus we use $x_d = 0$; the boundary layer then occurs at $x = 0$. Furthermore, it is easily seen that $a = 1$. Rewriting (3.1a) in the inner variable one finds

$$(1 + \tilde{x}) \frac{d^2 g}{d\tilde{x}^2} + \frac{dg}{d\tilde{x}} = \epsilon , \quad \epsilon \tilde{x} = x , \quad u(x, \epsilon) = g(\tilde{x}, \epsilon) . \tag{3.3}$$

Due to the linearity of the equation (cf. Section 5) we shall assume Poincaré-type forms of the inner and outer expansions, with the reservation that we may later have to number and possibly group the terms, in a sense which will be explained later. Thus the expansions have the form

$$\text{Outer}: \quad u \sim \sum_{j=0} \mu_j(\epsilon) \, f_j(x) , \tag{3.4a}$$

$$\text{Inner}: \quad u \sim \sum_{j=0} \nu_j(\epsilon) \, g_j(\tilde{x}) . \tag{3.4b}$$

The determination of the asymptotic sequences $\{\mu_j\}$ and $\{\nu_j\}$ will be part of the problem.

Since the boundary layer occurs at $x = 0$, f_0 has to satisfy the outer condition $f_0(1) = 2$. This suggests

$$\mu_0 = 1 . \tag{3.5}$$

(Any other nonzero constant would have been acceptable but we choose the simplest one.) The equation for f_0 (the reduced equation) is obtained by putting $\epsilon = 0$ in (3.1a), and the solution satisfying (3.1c) is

$$f_0(x) = x + 1 + C_0 \log x . \qquad (3.6)$$

Note that the outer equations will be of second order and there is only one boundary condition for each f_j,

$$f_0(1) = 2 , \quad f_j(1) = 0 \text{ for } j > 0 . \qquad (3.7)$$

Hence, unlike previously treated cases,[†] the outer expansion cannot be determined independently of the inner expansion (even if the μ_j were known). There will be constants C_j to be determined by matching.

The equation for g_0 is not known *a priori*. The term ϵ of the right-hand side disappears if $\epsilon \prec \nu_0$ and remains if $\nu_0 = \epsilon$ ($\nu_0 \prec \epsilon$ gives a nonsensical equation). Hence one finds two possible cases:

$$g_0 = D_0 \log(1 + \tilde{x}) , \quad \epsilon \prec \nu_0 , \qquad (3.8a)$$

or

$$g_0 = x + D_0 \log(1 + \tilde{x}) , \quad \epsilon = \nu_0 . \qquad (3.8b)$$

The function f_0 should match with g_0 to order unity. The choice (3.8b) gives

$$f_0 - \epsilon g_0 = \left[(x + 1 + C_0 \log x) - \left(x + \epsilon D_0 \log\left(1 + \frac{x}{\epsilon}\right) \right) \right] .$$

Here the constant term in f_0 cannot be matched. Thus $\epsilon \prec \nu_0$ and one finds

$$f_0 - \nu_0 g_0 = (x + 1 + C_0 \log x) - \nu_0 \log\left(1 + \frac{x}{\epsilon}\right) . \qquad (3.9)$$

[†] As we have seen there is a mild correction to this statement for (1.44). The outer expansion there has an additive constant which was determined by matching.

Since only $D_0 \nu_0$ matters we have given D_0 a simple value, namely $D_0 = 1$. For ϵ small, x fixed ($\bar{x}$ large),

$$g_0 = \log x - \log \epsilon + O_s\left(\frac{\epsilon}{x}\right) .$$

Thus putting

$$C_0 = 0 , \quad \nu_0 = -\frac{1}{\log \epsilon} , \tag{3.10a, b}$$

one finds that (3.9) tends to zero in a very small overlap domain. Hence matching determines not only the constant C_0 but also the order of the expansion parameter ν_0. The constant term in f_0 has been canceled identically by the choice of ν_0. The term x vanishes by itself if $x = o(1)$. However, $\nu_0 \log x$ must also tend to zero by itself which will not happen if x is too small. Putting $\eta x_\eta = x$ the overlap domain is

$$\left\{ \eta \mid -(\log \epsilon)^{-1} \prec -(\log \eta)^{-1} \text{ and } \eta \prec 1 \right\} . \tag{3.11}$$

An example of a function $\eta(\epsilon)$ in the overlap domain is

$$\eta \equiv \nu_0 = |\log \epsilon|^{-1} . \tag{3.12}$$

The function $\eta = \epsilon^a$ is not contained in this domain for any $a > 0$. We shall later (Section 5) encounter the expansion parameter ν_0 and an overlap domain similar to (3.11).

Whenever ν_0 is an expansion parameter one may replace it by a function of the same order

$$\nu_0^* = -\frac{1}{\log \epsilon + \delta(\epsilon)} , \quad \delta(\epsilon) \prec \log \epsilon . \tag{3.13}$$

It will be shown later that in the present case, and in other cases, a suitable choice of $\delta(\epsilon)$ may simplify the form of the expansion and increase its numerical accuracy.

The equation for f_1 is

$$x \frac{d^2 f_1}{dx^2} + \frac{df_1}{dx} = \begin{cases} 0 & \text{if } \mu_1 \succ \epsilon , \\ -\frac{d^2 f_0}{dx^2} & \text{if } \mu_1 = \epsilon . \end{cases} \tag{3.14a, b}$$

(An important feature of the present problem is that the equations for f_j, $j > 0$, and for g_j cannot be determined *a priori* but depend on the orders of μ_j and ν_j.) In the first case the solution is

$$f_1 = \log x \,, \tag{3.15}$$

where again the multiplicative constant of integration is absorbed in μ_1. We shall try to match to higher order without considering g_1; for this we form the difference

$$f_0 + \mu_1 f_1 - \nu_0 g_0 = [(x + 1) + \mu_1 \log x] - \nu_0 \left[\log x - \log \epsilon + \log \left(1 + \frac{\epsilon}{x}\right) \right] \,. \tag{3.16}$$

Here $\nu_0 \log \epsilon$ cancels the constant 1. The term $\nu_0 \log x$ is canceled if one puts

$$\mu_1 = \nu_0 \,. \tag{3.17}$$

In matching f_0 and $\nu_0 g_0$ we used the gauge function

$$\varsigma_0 = 1 \,. \tag{3.18}$$

A suitable gauge function for (3.16) is

$$\varsigma_1 = \nu_0 \,. \tag{3.19}$$

(Gauge functions are not unique.) In this case the overlap domain is

$$\left\{ \eta \,\middle|\, \epsilon \prec \eta \prec -\frac{1}{\log \epsilon} \right\} \,. \tag{3.20}$$

We note in passing that this domain has no function in common with the previous domain; the two overlap domains do not overlap. In some sense (which we find useless to define) the second overlap domain is larger.

Having found that $\mu_1 = \nu_0$ works successfully we abandon the choice $\mu_1 = \epsilon$ and (3.14b). In fact, it may be verified that this choice would not work.

The equation for g_1 again has one of two possible forms:

$$(1 + x)\frac{d^2 g_1}{d\tilde{x}^2} + \frac{dg_1}{d\tilde{x}} = \begin{cases} 0 & \text{if } \nu_1 \prec \epsilon \,, \\ 1 & \text{if } \nu_1 = \epsilon \,. \end{cases} \tag{3.21a,b}$$

The second alternative looks promising since eventually the term x in (3.6) has to be canceled by a term in the inner expansion. The solution of (3.21b) is

$$g_1(\tilde{x}) = D_1 \log(1 + \tilde{x}) + \tilde{x} . \tag{3.22}$$

Then

$$(f_0 + \nu_0 f_1) - (\nu_0 g_0 + \epsilon g_1) = -\nu_0 \log\left(1 + \frac{\epsilon}{x}\right) - \epsilon D_1 \log\left(1 + \frac{x}{\epsilon}\right) ,$$

which will go to zero when divided by the gauge function ϵ in the overlap domain (3.20) if $D_1 = 0$. Further calculation gives

$$\text{Outer expansion :} \quad \nu_0 = \frac{-1}{\log \epsilon} ,$$

$$u \sim (1 + x) + \nu_0 \log x + \epsilon \nu_0 \left(\frac{1}{x} - 1\right) \tag{3.23a}$$

$$+ \epsilon \nu_0^2(-\log x) + \epsilon^2 \nu_0 \left(\frac{1}{2} - \frac{1}{2x^2}\right) + \cdots .$$

Inner expansion :

$$u \sim \nu_0 \log(1 + \tilde{x}) + \epsilon \tilde{x} + \epsilon \nu_0^2 [-\log(1 + \tilde{x})] \tag{3.23b}$$

$$+ \epsilon^2 \nu_0^2 \frac{1}{2} \log(1 + \tilde{x}) + \cdots .$$

The beginning of a composite expansion is

$$u \sim x + \left[-\frac{1}{\log \epsilon} - \frac{\epsilon}{(\log \epsilon)^2} + \frac{1}{2}\frac{\epsilon^2}{(\log \epsilon)^2}\right] \log\left(1 + \frac{x}{\epsilon}\right) . \tag{3.24}$$

Improvement of the expansion. While asymptotically correct, the expansion given above is unsatisfactory for numerical purposes because the "small" parameter ν_0 becomes infinite when $\epsilon = 1$. Its numerical accuracy for small but finite ϵ may be improved by a technique introduced by Kaplun in 1956. This method has been called "telescoping" by Van Dyke (1975, p. 241). Recalling the freedom of choosing $\nu^*(\epsilon)$ defined by (3.13) one notes that the coefficient of $\log(1 + \tilde{x})$ in the composite expansion is to the order considered equivalent to

$$\nu_0^* = \frac{-1}{\log \epsilon - \epsilon + \frac{1}{2}\epsilon^2} . \tag{3.25}$$

The composite expansion then simplifies to

$$f \sim x + \nu_0^*(\epsilon) \log\left(1 + \frac{x}{\epsilon}\right) . \tag{3.26}$$

The second two terms in the denominator of ν_0^* are the beginning of the power series expansion of $\log(1 + \epsilon)$. Computation of higher order terms would have verified this. Then further telescoping can be done by replacing ν_0^* by

$$\nu_0^{**} = \frac{1}{-\log \epsilon + \log(1 + \epsilon)} . \tag{3.27}$$

For this simple problem the exact solution has thus been found by perturbation techniques.

The problem treated in the present section has several new distinctive features. The expansion parameters are not integral powers of the scaling parameter ϵ. They are not known *a priori* and, as a consequence, the equations for $f_j(x)$ and $g_j(x)$ are not known *a priori*. The expansion parameters are determined in the course of matching, together with certain unknown constants. Furthermore $\mu_j \neq \nu_j$; in fact their orders are different. While the expansions are of Poincaré form and hence may be obtained by repeated application of limit processes, their numbering (ordering) must be observed. One finds $\lim_\epsilon u = 0$ but it does not make sense to call this the first term of the inner expansion. The first term is (see, however, comments on grouping below),

$$\nu_0 \lim_\epsilon \left(u \nu_0^{-1}\right) = \nu_0 \log(1 + \tilde{x}) . \tag{3.28}$$

This is the term which matches to order unity with f_0.

Remarks on Numbering and Grouping of Terms. If one expands $\nu_0 g_0$ in outer variables one finds

$$\nu_0 \log\left(1 + \frac{x}{\epsilon}\right) = 1 + \nu_0 \log x + \frac{\epsilon \nu_0}{x} - \frac{\epsilon^2 \nu_0}{2x^2} \cdots . \tag{3.29}$$

Thus the common part with f_0 is 1 and the composite expansion is

$$C^{(0)}[u] = x + \nu_0 \log \left(1 + \frac{x}{\epsilon}\right) . \tag{3.30}$$

However (3.30) also contains $\mu_1 f_1 = \nu_0 \log x$. Thus whether one uses one or two terms of the outer expansion one obtains the same composite expansion. Furthermore, $\nu_1 g_1 = \epsilon \tilde{x} = x$ is contained in $\mu_0 f_0 = 1 + x$. Taking the second term of the inner expansion into account still gives the composite expansion (3.30). This leads to the question: should one group the terms, for instance by considering $(1 + x) + \nu_0 \log x$ as the first term of the outer expansion? In the present case such a decision is rather arbitrary. The main objective fact is the statement about the composite expansion just made. Another fact is that in (3.6) one could have put $C_0 = \nu_0$ rather than zero and thus have obtained $\mu_0 f_0 + \mu_1 f_1$ directly by matching with $\nu_0 g_0$. Thus grouping is possible, but whether it should be done or not seems rather arbitrary. A consequence is the ambiguity in the definitions of $E^{(k)}[u]$ and $H^{(k)}[u]$.

Second Example.[†] *Techniques of Matching Compared with Underlying Ideas.* In the discussion to follow there is less stress on the example given than on general ideas. We are concerned less with developing new techniques than with the interpretation of existing techniques. In principle the discussion could have been put in Chapter I; however, it seems more appropriate to put it here after the specific examples given in previous sections. We are concerned with the role of *overlap* in matching and, in discussing this, we shall have to reconsider the question of *correction* layers. In many examples the first term, or several terms, of the outer expansion can be found independent of the inner expansion. We then form the difference function (to a certain order) by subtracting a known partial outer expansion from the (unknown)

[†] The discussion here is based on Lagerstrom (1976). See also Van Dyke (1975), Notes 4 and 5.

exact solution. The equations and the boundary conditions of the difference function are easily found. Applying layer technique one finds that the outer expansion, to the order considered, is identically zero. The inner expansion obtained then describes the correction layer. (Equation (1.10) and following equations illustrate this.) We claim that no general statements can be made about the usefulness of this concept. In the case to be discussed here correction layers are essential for understanding the matching. In other cases they are practical (see Section 1): Terms which one feels really belong to the inner expansion are put there and do not also appear in the outer expansion. Finally, we shall see in Section 5 that there are examples for which the correction layer is useless.

Our main purpose here is to discuss the idea of matching. Matching involves a comparison of a partial outer expansion and a partial inner expansion; let us call them $E^{(m)}[u]$ and $H^{(n)}[u]$ (here m and n may or may not be the same integers). It seems to us intuitively clear that for comparison of two expressions there must be a domain where both expressions are valid, that is, a domain of overlap. We repeat here a description of the matching technique used. One forms the difference between the two partial expansions and divides by a gauge function $\varsigma(\epsilon)$. This expression should tend to zero in the domain of overlap. This means that 1) certain terms cancel others automatically; 2) other terms cancel because they are small compared to $\varsigma(\epsilon)$ in the domain considered; 3) finally, certain terms cancel identically only if certain previously unknown constants are given certain values. The third item is the essence of determining undetermined constants by matching. To check the second item one may introduce a variable x_η by $x = \eta(\epsilon)x_\eta$ and let $\epsilon \downarrow 0$ with x_η fixed. This is the *principle* of intermediate matching or matching by overlap; here $\eta(\epsilon)$ belongs to the overlap domain (actually the overlap domain may be guessed at so to speak, experimentally, by finding the $\eta(\epsilon)$ for which the second step works). We recommend the explicit use of an intermediate variable x_η

only for two purposes: 1) For pedagogical purposes in a few simple examples in order to show the underlying idea and 2) for intricate problems (a very subjective description) which need a very careful analysis. However, *in practice* it is usually sufficient simply to use outer variables and estimate directly which terms cancel for some appropriate order of x.

Fraenkel (1969, pp. 216–217) gives an example for which he claims that expansions which do not overlap still may be matched by the formal matching method (see (1.13) and the discussion of (1.16)). We shall discuss this idea here, making assumptions which apply to Fraenkel's example (but not in general). We assume first that $E^{(k)}[u]$ and $H^{(k)}[u]$ and also $E^{(k)}H^{(k)}[u]$ and $H^{(k)}E^{(k)}[u]$ can be properly defined. We also assume the equality (commutativity)

$$H^{(k)}E^{(k)}[u] = E^{(k)}H^{(k)}[u] \; , \tag{3.31}$$

and that the composite expansion defined by

$$C^{(k)}[u] = \left(E^{(k)} + H^{(k)} - E^{(k)}H^{(k)} \right)[u] \; , \tag{3.32}$$

is uniformly valid to order $\varsigma_k(\epsilon)$, that is,

$$u - C^{(k)}[u] = o(\varsigma_k) \; , \tag{3.33a}$$

in the entire interval considered. We define the difference function $D^{(k)}[u]$ by

$$v = D^{(k)}[u] = u - E^{(k)}[u] \; , \tag{3.33b}$$

and rewrite (3.33b), using (3.31), as

$$v - H^{(k)}[v] = o(\varsigma_k) \; ; \tag{3.34a}$$

since $E^{(k)}E^{(k)}[u] = E^{(k)}[u]$ it follows that $E^{(k)}[v] = 0$ and hence

$$v - E^{(k)}[v] = u - E^{(k)}[u] \; . \tag{3.34b}$$

Thus if $E^{(k)}[u]$ and $H^{(k)}[u]$ are both valid to order ς_k, possibly in a non-overlapping domain, one finds that under the assumptions stated $H^{(k)}[v]$ is an approximation to v valid to order $\varsigma_k(\epsilon)$ in the entire domain (uniformly) and since $E^{(k)}[v]$ is valid in some domain the two approximations to v have an overlap domain.

Rather than discussing obvious generalizations of the reasoning given above, we consider a concrete example. We choose a simple form of the function given by Fraenkel (1969),

$$u(x,\epsilon) = U(\tilde{x})V(x),\ U(\tilde{x}) = \sum_{j=0}^{N} \tilde{x}^{-j},\ V(x) = \sum_{j=0}^{N} x^j,\ \epsilon\tilde{x} = x\ . \tag{3.35}$$

The x-domain is $[\epsilon, 1]$ and N should be large enough to make the problem interesting. If one takes $N = 3$ the complete outer expansion is

$$u = \sum_{j=0}^{3} x^j + \epsilon \sum_{j=-1}^{2} x^j + \epsilon^2 \sum_{j=-2}^{1} x^j + \epsilon^3 \sum_{j=-3}^{0} x^j\ , \tag{3.36a}$$

and the complete inner expansion is

$$u = \sum_{j=-3}^{0} \tilde{x}^j + \epsilon \sum_{j=-2}^{1} \tilde{x}^j + \epsilon^2 \sum_{j=-1}^{2} \tilde{x}^j + \epsilon^3 \sum_{j=0}^{3} \tilde{x}^j\ . \tag{3.36b}$$

For simplicity we shall only consider gauge functions ϵ^a, $a \geq 0$ and denote the limit in which $x\epsilon^{-a}$ is fixed by $\lim_a$. One finds for $a, b \geq 0$,[†]

$$\lim_a \frac{u - E^{(p)}}{\epsilon^b} = 0 \text{ iff } b < (1+p)(1-a)\ , \tag{3.37a}$$

$$\lim_a \frac{u - H^{(p)}}{\epsilon^b} = 0 \text{ iff } b < (1+p)a\ . \tag{3.37b}$$

In Figure 3.1 we plot the curves $b = (1+p)(1-a)$ and $b = (1+p)a$. Calling I_p the intersection for a given p the coordinates of this point are $a = .5$, $b = \frac{1}{2}(p + 1)$.

[†] In the present case, with simple expansion parameters, there is no ambiguity in defining $H^{(k)}$ and $E^{(k)}$.

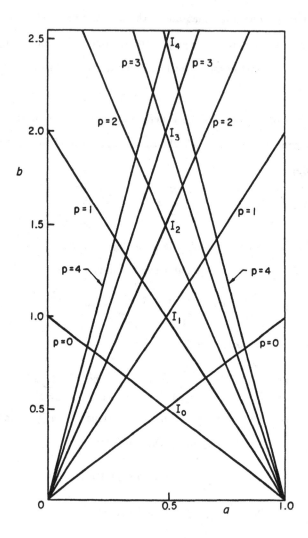

Figure 3.1. Graphs of $b = (1+p)a$, and $b = (1+p)(1-a)$; $p = 0, 1, 2, 3, 4$.
Two curves with the same value of p intersect at $a = .5$,
$b = (1+p)/2$. We call this point I_p.

$E^{(p)}$ and $H^{(p)}$ then overlap in the triangle with apex at I_p and the lateral sides excluded. The overlap with respect to the gauge function ϵ^b is the open interval between the two values of a where the line $b = $ constant intersects the left and right leg of the corresponding triangle. As usual the overlap domain shrinks as the order of the gauge function increases. We notice that $E^{(2)}$ and $H^{(2)}$ overlap only for gauge functions ϵ^b, $b < 1.5$; $H^{(3)}$ and $E^{(3)}$ overlap only for gauge functions ϵ^b, $b < 2$ etc. We now consider overlap of the difference function for $p = 2$,

$$v = D^{(2)}[u] = u - E^{(2)}[u] = \epsilon^3 \left(\frac{1}{x^3} + \frac{1}{x^2} + \frac{1}{x} + 1 \right) , \qquad (3.38a)$$

$$E^{(2)}[v] = 0 , \; H^{(2)}[v] = \frac{1}{\tilde{x}^3} + \frac{\epsilon}{\tilde{x}^2} + \frac{\epsilon^2}{\tilde{x}} = v - \epsilon^3 , \qquad (3.38b)$$

$$\frac{H^{(2)}[v] - E^{(2)}[v]}{\epsilon^b} = \epsilon^{3-b} \left(\frac{1}{x^3} + \frac{1}{x^2} + \frac{1}{x} \right) . \qquad (3.38c)$$

Thus choosing as an example $b = 2$, we see that while $E^{(2)}[u]$ and $H^{(2)}[u]$ do not overlap to order ϵ^2, the expansions $E^{(2)}[v]$ and $H^{(2)}[v]$ do overlap to order ϵ^2 in any x-interval such that $\epsilon^{1/3} \prec x \leq 1$. Thus by studying a simplified version of Fraenkel's example we have seen that his objections to intermediate matching can be avoided by using difference functions.

However, a main purpose of matching is to determine constants of integration in the asymptotic solution of a differential equation. We must therefore show that by using a difference function as in (3.38) we can determine such constants by matching, without using higher order terms of $E^{(p)}[u]$ and $H^{(p)}[u]$. We shall therefore study a differential equation related to the function discussed above.

We observe that for $N = 2$, $u(x, \epsilon)$ satisfies the equation

$$\epsilon \frac{d^2 u}{dx^2} + \frac{du}{dx} = (1 + 2x) + \epsilon(-\frac{1}{x^2} + 3)$$
$$+ \epsilon^2(-\frac{1}{x^2}) + \epsilon^3(\frac{6}{x^4} + \frac{2}{x^3}) . \qquad (3.39a)$$

As boundary conditions we take

$$u(1,\epsilon) = 3 + \epsilon(3 + a_1) + \epsilon^2(3 + a_2) , \qquad (3.39b)$$

$$u(\epsilon,\epsilon) = 3 + \epsilon(3 + b_1) + \epsilon^2(3 + b_2) . \qquad (3.39c)$$

The outer expansion and inner expansions will be denoted by

$$u(x,\epsilon) = \sum_{j=0} \epsilon^j f_j(x) , \qquad (3.40a)$$

$$u(x,\epsilon) = g(\tilde{x},\epsilon) = \sum_{j=0} \epsilon^j g_j(\tilde{x}) . \qquad (3.40b)$$

One finds immediately

$$f_0 = 1 + x + x^2 , \qquad f_1 = \frac{1}{x} + x + 1 + a_1 , \qquad (3.41a,b)$$

$$f_2 = \frac{1}{x^2} + \frac{1}{x} + 1 + a_2 , \qquad f_j = 0 \text{ for } j > 2 . \qquad (3.41c,d)$$

The function g_0 is obtained in a straightforward way by using the boundary condition at $x = \epsilon$ and matching with gauge function unity:

$$g_0 = \tilde{x}^{-2} + \tilde{x}^{-1} + 1 . \qquad (3.42)$$

Using the inner boundary condition one finds

$$g_1 = \tilde{x} + 1 + \tilde{x}^{-1} + C_1 + D_1 e^{-\tilde{x}} , \quad C_1 + D_1 e^{-1} = b_1 ; \qquad (3.43)$$

thus

$$E^{(1)}[u] - H^{(1)}[u] = x^2 + \epsilon x + \epsilon a_1 - \frac{\epsilon^2}{x^2} - \frac{\epsilon^2}{x} - \epsilon C_1 + \text{TST} . \qquad (3.44)$$

Dividing by ϵ^b and writing $x = \epsilon^a x_a$ one finds that the various terms are multiplied by the following powers of ϵ:

$$\epsilon^{2a-b}, \quad \epsilon^{1+a-b}, \quad \epsilon^{2-2a-b}, \quad \epsilon^{2-a-b}, \quad \epsilon^{1-b} . \qquad (3.45)$$

The first three tend to zero if

$$\frac{b}{2} < a < 1 - \frac{b}{2} , \tag{3.46}$$

which presupposes $b < 1$ (cf. Figure 3.1). Then of course $\epsilon^{1-b}(a_1 - C_1)$ will tend to zero. However, we still conclude

$$a_1 = C_1 , \tag{3.47}$$

without computing higher order terms. It is sufficient to observe that if constant terms occur in f_k and g_k, $k > 1$, they will be multiplied by ϵ^k and hence cannot cancel $\epsilon(a_1 - C_1)$. Thus for matching $E^{(1)}[u]$ and $H^{(1)}[u]$ we have used a gauge function which is $\prec \epsilon$ while ϵ is the highest-order expansion parameter; the possibility of this was pointed out to the author by J. Boa.

Coming back to the original problem it is easily seen that straightforward matching can easily be applied to the difference functions $v = D^{(k)}[u]$. This is left as an exercise for the reader.

Discussion. In previous sections we demonstrated how one may systematically find suitable *scaling factors* for forming the inner variables. In the first example above we encountered new types of *expansion coefficients*; it was shown how these may be determined by a rational procedure. The expansions are not unique and it was shown how this ambiguity may be advantageous. In particular we found coefficients involving $\log \epsilon$. In subsequent sections (for instance Sections 5 and 6) it will be seen that such coefficients are not freaks occurring for artificial problems; such coefficients occur naturally in physical problems. The question about numbering and grouping the terms also came up; it was seen that sums of two terms with different expansion coefficients in some sense should be counted as one; they belong together. This phenomenon will also occur in subsequent sections.

The second example illustrated how matching by overlap is the underlying principle even for matching partial expansions which do not seem to overlap. This problem was connected with the use of correction layers, a subject which will be further discussed in Section 5.

2.4. A Simple Nonlinear Equation Exhibiting a Wide Variety of Solutions

Introduction. We now resume the trend of §1 and §2 by a further generalization of Friedrichs' example: The coefficient of u' now depends on u, so that the problem is quasilinear. We choose a simple-looking problem[†] which, however, will turn out to have many types of solutions,

$$\epsilon\, u'' + uu' - u = 0 \;,\quad u(0) = A \;,\quad u(1) = B \;. \qquad (4.1a,b,c)$$

For simplicity we assume that A and B are independent of ϵ. Since the equation is autonomous solutions given below trivially lead to solutions for any displaced unit interval. A change of the length of the interval does not introduce any significant changes, provided the interval is finite.

[†] The problem was introduced by the author in an informal discussion group on singular perturbations at Caltech in the 1950s. The goal was not to study nonlinearity but to find a simple model for shock layers in fluids by analogy with Friedrichs' model of boundary layers. The introduction of the term uu' then was natural; by analogy with fluid dynamics expected solutions with shock layers (interior layers) or boundary layers at either endpoint were found by standard means. Saul Kaplun pointed out a typical feature of nonlinearity, namely the possibility of scaling the dependent as well as the independent variable with functions of ϵ. He found an important class of asymptotic solutions, here called derivative layers. His scaling gave an ϵ-independent form of the equation. By using a first integral J. D. Cole then could give a universal phase portrait (Figure 4.1), and by utilizing an additional type of inner solution, boundary layers with algebraic decay, he gave a complete picture of the (A, B)-plane (Figure 4.2) which neatly illustrated the various types of layers necessary in various domains. The results, including Figures 4.1 and 4.2, were first published in Cole (1968). Higher-order solutions were discussed in Lagerstrom (1976), and in Kevorkian-Cole (1981). For a discussion of general classes of equations having the present problem as a special case, see Chang and Howes (1984). Their treatment stresses mathematical rigor and generality and contains many references.

Outer Expansion. As usual, f_0, the leading term of the outer expansion, obeys the reduced equation

$$f_0 f_0' - f_0 = 0 . \tag{4.2}$$

This has two different types of solutions,

$$f_0^{(1)} = x + C , \quad f_0^{(2)} = 0 , \tag{4.3a, b}$$

where C is a constant. We shall see that, for a range of values of A and B, $f_0^{(1)}$ is discontinuous at either boundary point or at an interior point. The interior discontinuity may be smoothed out by an interior layer of thickness ϵ provided the jump is symmetric around zero, that is it goes from $-a$ to $+a$, $a > 0$. We may also piece together the functions $f_0^{(1)}$ and $f_0^{(2)}$ to obtain a continuous function which then necessarily has a jump in the derivative which can be smoothed out by a derivative layer. An example is $f_0 = 0$ in $[0, x_d]$ and $f_0 = x - x_d$ in $[x_d, 1]$ with $0 < x_d < 1$. While $f_0^{(1)}$ and $f_0^{(2)}$ are individually exact solutions of $(4.1a)$ the solution pieced together as above is not. Kaplun used his scaling to show the asymptotic significance of such solutions.

Higher-order terms in the outer expansion are identically zero since $f_0'' = 0$.

Layers of Thickness ϵ. If we use a function of ϵ to scale the independent (but not the dependent) variable to obtain a suitable inner equation we find that the scaling must be $O_s(\epsilon)$ to obtain the richest equation. Therefore we define the inner variable $\tilde{x}$ by

$$\epsilon \, \tilde{x} = x - x_d , \tag{4.4}$$

which gives the exact equation

$$\frac{d^2 g}{d\tilde{x}^2} + g \, \frac{dg}{d\tilde{x}} - \epsilon g = 0 , \quad g(\tilde{x}, \epsilon) = u(x, \epsilon) . \tag{4.5a, b}$$

The leading term of the inner expansion then satisfies[†]

$$\frac{d^2 g_0}{d\tilde{x}^2} + g_0 \frac{dg_0}{d\tilde{x}} = 0 , \tag{4.6}$$

whose nonconstant solutions are, with b and $k =$ constants,

$$g_0 = b \tanh \left[\frac{b}{2}(\tilde{x} + k)\right] , \quad g_0 = b \coth \left[\frac{b}{2}(\tilde{x} + k)\right] , \tag{4.7a, b}$$

$$g_0 = \frac{2}{\tilde{x} + k} , \tag{4.7c}$$

$$g_0 = -b \tan \left[\frac{b}{2}(\tilde{x} + k)\right] , \quad g_0 = b \cot \left[\frac{b}{2}(\tilde{x} + k)\right] . \tag{4.7d, e}$$

Since all solutions are odd in $\tilde{x} + k$ we may assume $b > 0$. Actually, all solutions above are easily derivable from the special case $g_0 = \tanh(\tilde{x}/2)$. We notice first that the equation is invariant[‡] under the mappings

$$\tilde{x} \to \tilde{x} + k, \quad g_0 \to g_0 \text{ (translational invariance due to autonomy)} , \tag{4.8a}$$

$$\tilde{x} \to b\tilde{x}, \quad g_0 \to b^{-1} g_0 . \tag{4.8b}$$

From the invariances of (4.6) it follows that

$$g_0(\tilde{x}) \text{ a solution} \implies b \, g_0\big[b(\tilde{x} + k)\big] \text{ a solution.} \tag{4.9}$$

Starting from $\tanh(\tilde{x}/2)$ we obtain $-\tan(\tilde{x}/2)$ by putting $b = i$, $k = 0$. From this we find $\cot(\tilde{x}/2)$ by replacing $\tilde{x}$ by $\tilde{x} + \pi$ and then $\coth(\tilde{x}/2)$ by using $b = i$.[††] Finally, as $b \to 0$, (4.7b) converges to (4.7c); the convergence is not uniform in $\tilde{x}$.

[†] (4.6) is well known from gas dynamics. It was introduced by G. I. Taylor for a simple description of the structure of a shock layer.

[‡] The full equation (4.1a) is obviously invariant under (4.8a) but not under (4.8b). It actually has, in principle, an invariance generalizing (4.8b) which has not been given in explicit form and has, so far, not been of any practical use.

[††] We may also obtain $\coth(\tilde{x}/2)$ from $\tanh(\tilde{x}/2)$ by replacing $\tilde{x}$ by $\tilde{x} + i\pi$.

To find the possible use of (4.7) as layer solutions we observe that the hyperbolic tangent as given by (4.7a) asymptotes to $-b$ at $-\infty$ and to b at $+\infty$ and is monotone increasing. It may therefore be used as an interior layer (to smooth out a jump from $-b$ to b, $b > 0$, in the outer solution), and as an increasing boundary layer at $x = 0$ or a decreasing (for $\tilde{x}$ decreasing) boundary layer at $x = 1$. The hyperbolic cotangent has two branches, a negative decreasing one for negative arguments which asymptotes to $-b < 0$ as $x \to -\infty$. As the argument increases to zero it tends to $-\infty$, then jumps from $-\infty$ to $+\infty$ of the positive branch which is monotone decreasing to the limiting value $b > 0$ as $x \to +\infty$. It can therefore not be used as an interior layer (which must be two-sided). It may be used as a boundary layer at either endpoint in the way the hyperbolic tangent is used (except that the former is decreasing when the latter is increasing).

To illustrate the situation concretely we fix the value of B to be $B = 3$ and study the various cases as A decreases from $+\infty$ to $-\infty$.

1) $A > 2$: The boundary layer is at $x = 0$. The outer solution is

$$f_0 = x + 2 , \tag{4.10}$$

and the layer solution

$$g_0 = 2\coth(\tilde{x} + k) , \quad x_d = 0 , \quad A = 2\coth k . \tag{4.11a, b, c}$$

Here we use the positive branch of the hyperbolic cotangent.

2) $-2 < A < 2$: At $A = 2$ there is no layer at all and the outer solution (4.10) is exact. As A decreases we get f_0 as in (4.10) and

$$g_0 = 2\tanh(\tilde{x} + k) , \quad x_d = 0 , \quad A = 2\tanh k . \tag{4.12a, b, c}$$

3) $-4 < A < -2$: For $A < -2$, (4.12c) cannot be solved. The boundary layer then detaches itself and becomes an interior layer which drifts to the

right as A decreases. There are then two pieces of the outer solution which is discontinuous at $x = x_d$,

$$f_{OL} = x + A , \quad 0 \le x < x_d , \tag{4.13a}$$

$$f_{OR} = x + 2 , \quad x_d < x \le 1 . \tag{4.13b}$$

Both boundary conditions are satisfied. The discontinuity must be smoothed out by a two-sided layer that is a hyperbolic tangent which asymptotes to $\pm b$. Thus $-f_{OL}(x_d) = f_{OR}(x_d) = b$, which implies

$$x_d = -1 - \frac{A}{2} , \quad b = 1 - \frac{A}{2} , \quad k = 0 . \tag{4.14a, b, c}$$

Note that for a two-sided layer the discontinuity occurs when the argument of the hyperbolic tangent is zero. This explains the choice of k and incidentally gives $f_0(x_d) = 0$, that is, the mean value of the limiting values of f_{OL} and f_{OR} as x approaches x_d.

4) $A < -4$: For $A < -4$, (4.14a) gives $x_d > 1$ which is outside the domain considered. The layer has now attached itself to the right endpoint. Then $f_0 = f_{OL}$ as given above is the entire outer solution and the layer is now

$$g_0(\tilde{x}) = -(1 + A) \tanh\left[-\frac{(1 + A)}{2} (\tilde{x} + k)\right] , \tag{4.15a}$$

where

$$x_d = 1 , \tag{4.15b}$$

and k is determined by

$$3 = -(1 + A) \tanh\left(-\frac{(1 + A)(1 + k)}{2}\right) . \tag{4.15c}$$

Note that for $A < -4$ the positive constant $-(1 + A)$ is always > 3 so that (4.15c) always has a unique solution.

Comments.

a) The fact that b always had the value 2 in Cases 1 and 2 depends of course on the fact that $f(1)$ remained constant $= 3$ and that the boundary layer had to approach 2 whenever it was located at $x = 0$. In those two cases the choice $x_d = 0$ is correct. Applying the outer limit to $g_0(\tilde{x})$ one gets the value 2 in $(0,1]$ and A at $x = 0$. The discontinuity thus is located at $x = 0$. A similar argument shows that $x_d = 1$ is correct in Case 4 although, since A varies, the value of b is not fixed. This asymmetry has no significance; it is due to the arbitrary decision of fixing B and varying A.

b) A composite expansion, uniformly valid to order unity in $[0,1]$ is formed in the standard manner, as used in Section 1. For instance, in Case 1

$$C^{(0)}[f] = x + 2\coth(\tilde{x} + k) . \tag{4.16}$$

As usual, this is obtained by subtracting out the common part of f_0 and g_0. Actually, a similar trick works for Case 3. Consider

$$C^{(0)}[f] = b\tanh\frac{b\tilde{x}}{2} + x - x_d = \left(1 - \frac{A}{2}\right)\tanh\left(\frac{1}{2} - \frac{A}{4}\right)\tilde{x} + x + 1 + \frac{A}{2} . \tag{4.17}$$

This satisfies the correct boundary conditions within exponentially small terms and its outer limit is correct. In taking its inner limit one replaces $(x - x_d)$ by $\epsilon\tilde{x}$ so that its inner limit is also seen to be correct.

It will be seen later that if the next term in the inner expansion is found, then $C^{(1)}[f]$ is given by $g_0 + \epsilon g_1$.

c) The different regimes given above are open intervals in A. The limiting value $A = 2$ was discussed. The problem of the limiting values $A = -2$ and $A = -4$ is relegated to Exercise 4.3.

d) We have not used all possibilities provided by (4.7). The trigonometric functions (4.7d, e) are obviously useless; the other functions will be mentioned below.

There is, however, a serious lacuna in our scheme. There are regions of the (A, B)-plane for which the layers (4.7) are insufficient, as will now be shown.

Let us first consider what happens when B decreases to $B = 1$ while, say, $A > 0$. The hyperbolic cotangent describing the layer must then be replaced by the limiting case

$$g_0(\tilde{x}) = \frac{2}{\tilde{x} + 2/A} \cdot \qquad (4.18a)$$

Now let B decrease further and $0 < B < 1$. Then none of the previous combinations of layer solutions and outer solutions can describe the situation. Instead we propose tentatively the following outer solution:

$$f_0(x) = 0 \text{ in } [0, x_d] , \quad f_0 = x + B - 1 \text{ in } [x_d, 1] , \quad x_d = 1 - B , \qquad (4.18b)$$

keeping (4.18a) as the inner solution. This outer solution satisfies the boundary condition at $x = 1$ and matches with the inner solution since $g_0(\infty) = f_0(0) = 0$. In the previous constructions we showed that the outer solutions used were relevant solutions of the exact equation in the sense that they are outer limits of exact solutions (actually we showed that they were limits of a composite solution uniformly valid to order unity). The discontinuity could not be arbitrary; for instance an interior discontinuity has to go from $-|b|$ to $+|b|$ while a jump from -1 to $+2$ is not permissible in a relevant solution. We must therefore justify that the outer solution used above is a relevant solution. The solution was actually introduced by Kaplun who found it by a new scaling which turned out to be very useful for various aspects of the present problem (as well as for other nonlinear problems).

The Kaplun Scaling. Universal Phase Portrait. Derivative Layers. While for a linear equation any scaling of the dependent variable (whether by dimensional analysis or by multiplication by a function of a small parameter) usually does not give significant results, the situation is different for a nonlinear equation. Kaplun introduced the scaling

$$\epsilon^{1/2} h = u , \quad \epsilon^{1/2} \bar{x} = x - x_d . \tag{4.19a,b}$$

The resulting equation is

$$\frac{d^2 h}{d\bar{x}^2} + h \frac{dh}{d\bar{x}} - h = 0 , \tag{4.20}$$

which is the richest equation obtainable by a double scaling (see Figure 4.3 and Exercise 4.4). It happens to be the exact equation[†] rewritten in the new variables. We shall use this form to show that the outer solution (4.18b) is a relevant one.

First, however, we make some general observations about (4.20), which are of course, with trivial modifications, also valid for the original equation (4.1a). The equation is autonomous, hence it shares one symmetry in common with (4.6): If $h(\bar{x})$ is a solution then so is $h(\bar{x} + a), a = $ a constant. It is also invariant under a discrete symmetry

$$\bar{x} \rightarrow -\bar{x} , \quad h \rightarrow -h , \quad v \rightarrow v . \tag{4.21}$$

Here v is $dh/d\bar{x}$ and also du/dx. We write (4.20) as a system

$$\frac{dh}{d\bar{x}} = v , \quad \frac{dv}{d\bar{x}} = h(1 - v) , \tag{4.22a,b}$$

which leads to the integral[‡]

$$\frac{h^2}{2} + v + \log |1 - v| = C = \text{ constant} . \tag{4.23}$$

Obviously $v = 1$, or $h = \bar{x} + $ constant, is an exact solution represented by a horizontal line in the phase plane which no solution curve can cross. Another exact

---

[†] This is exceptional and depends on the fact that (4.1a) is made up of three monomials in u and its derivatives. Even when the full equation does not result, a similar scaling for the dependent variable of a nonlinear equation may be important.

[‡] This shows that (4.20) is in principle solvable by quadrature and inversion, although solutions cannot be written out explicitly (except those very special ones mentioned below).

solution is $h = v = 0$ which is a saddle point and the only singular point in the phase plane (see Figure 4.1).

As usual, curves leading into or going out of a singular point are of special interest. There is one such curve in each quadrant. The one in the nth quadrant will be denoted by $v = V_n(h)$, and corresponds to a solution $h = H_n(\bar{x})$. At the singular point $h = v = 0$ so that the special curves correspond to the value $C = 0$ in the integral (4.23).

We shall now discuss some relevant properties of the functions $V_n(h)$. We shall restrict ourselves to the first and fourth quadrants where $h \geq 0$; because of the discrete symmetry (4.21) this is sufficient.

First consider V_1. Since $1 - v > 0$ and $h > 0$, V_1 increases with h. On the other hand it cannot cross $v = 1$. It must therefore asymptote to the curve a value $v = v_0$. Since $h \uparrow \infty$ and $C = 0$ we deduce from (4.22b) or (4.23) that $v_0 = 1$. In the outer limit $\bar{x}$ tends to $+\infty$ for $x > x_d$ and since $V_1(\infty) = 1$ the outer limit is $f_0 = x - x_d$. When $x < x_d$ the outer limit of $\bar{x}$ is $-\infty$, that is, V_1 tends to the singular point. Hence for $x < x_d$ the outer limit is $V_1 = 0$ and the outer solution is $f_0 = 0$. Thus we have a solution which corresponds to the outer solution (4.18b) and $H_1(\bar{x})$ may be used to smooth out a discontinuity in the derivative of f_0. We therefore call H_1 a derivative layer.[†] While $H_1(\bar{x})$ cannot be expressed in an explicit tractable form, the exact integral for which $C = 0$ can give us much information about it. What matters most is that the same singular curve can be used for a large region of the (A, B)- plane (Figure 4.2). In that sense the functions V_1 and H_1 are universal.

Next we study V_4 and H_4. Since $v < 0$, the direction of the arrow for increasing

---

[†] Kaplun, who introduced the idea of this solution and its application, called it a corner layer or corner solution. Those terms are now used to describe an asymptotic solution valid near an actual corner of a boundary in a problem with several independent variables.

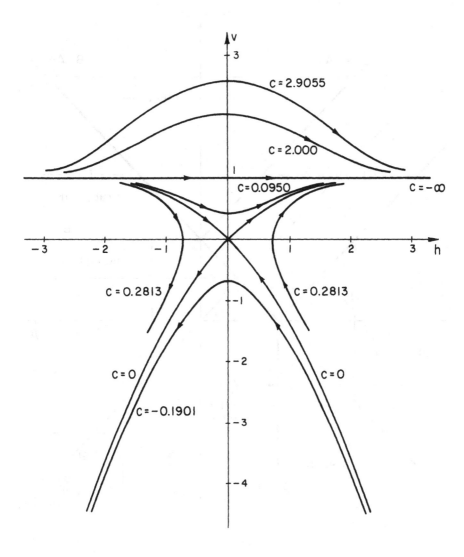

First integral: $\dfrac{h^2}{2} + v + \log |v - 1| = C$

$$\epsilon^{1/2} h = u, \quad \epsilon^{1/2} \bar{x} = x - x_d$$

Figure 4.1. Universal Phase Portrait. See (4.22) and (4.23).

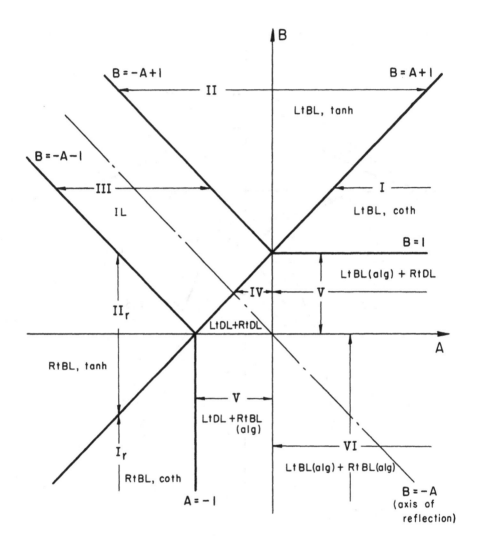

Notation: Lt = at left endpoint,   Rt = at right endpoint
           BL = boundary layer (exponential decay)
           BL(alg) = boundary layer (algebraic decay)
           IL = interior layer (always tanh)
           DL = derivative layer (thickness $\epsilon^{1/2}$)
           Subscript "r" indicates reflected region.

Figure 4.2. Dependence of Occurrence of Layers on $u(0) = A$, $u(1) = B$.

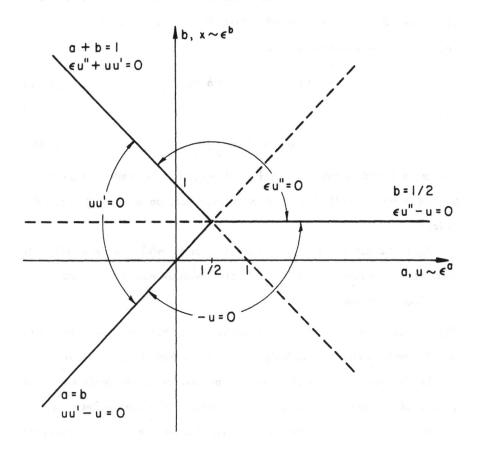

At $a = 1/2,$ $b = 1/2$ full equation is valid.

Figure 4.3. Limiting Equations for Different Scalings of x and u.

$\bar{x}$ is as given in Figure 4.1. There is no horizontal asymptote. As $|h|$ increases (decreasing $\bar{x}$) $|v|$ increases indefinitely. For $|v|$ large, $\log(v-1)$ may be neglected in comparison to v so that

$$v = V_4(h) \simeq -\frac{h^2}{2} \ , \ \text{as } h, \ |v| \to \infty \ , \qquad (4.24a)$$

which integrates to

$$h = H_4(\bar{x}) \simeq \frac{2}{\bar{x}+k} \ . \qquad (4.24b)$$

This is the boundary-layer with algebraic decay (4.7c). We leave it to the reader to show in detail that the inner limit of the exact solution $h = H_4(\bar{x})$ is actually (4.7c).[†]

Figure 4.2 shows which types of layers occur as $u(0) = A$ and $u(1) = B$ vary. This figure covers the entire (A, B)-plane with the exception of certain lines separating the regions.

Higher-Order Approximations. This problem is discussed by Lagerstrom (1976) and, for the boundary layer with algebraic decay, by Kevorkian-Cole (1981).

The higher-order terms of the outer expansion are all identically zero. The problem of the inner expansion remains, and we may also consider better approximations to the exact solutions $h = H_n(\bar{x})$. Assume that the inner expansion is

$$g(\tilde{x}, \epsilon) = g_0(\tilde{x}) + \epsilon g_1(\tilde{x}) + o(\epsilon) \ . \qquad (4.25)$$

Calculations will show us whether a switchback term (see Section 5) of the form $\eta(\epsilon)g_1(\tilde{x})$, $\epsilon \prec \eta \prec 1$, is needed. The equation for g_1 is

$$\frac{d^2 g_1}{d\tilde{x}^2} + \frac{d(g_0 g_1)}{d\tilde{x}} = g_0 \ , \qquad \epsilon\tilde{x} = x - x_d \ . \qquad (4.26)$$

[†] Cole, who introduced the use of this solution, called it a transition layer. We prefer the name "boundary layer with algebraic decay." This name should, strictly speaking, be applied to (4.7c), not to the exact function $h = H_4(\bar{x})$.

The integration of (4.26) is straightforward. Defining

$$G_0 = \int_0^{\tilde{x}} g_0(s)ds \ , \quad H_0 = e^{G_0} \ , \quad K_0 = \int_0^{\tilde{x}} H_0(s)ds \ , \qquad (4.27)$$

we find a particular solution

$$g_{1p} = H_0^{-1}(\tilde{x}) \int_0^{\tilde{x}} G_0(s)H_0(s)ds$$

$$= H_0^{-1}(\tilde{x}) \left[G_0(\tilde{x})K_0(\tilde{x}) - \int_0^{\tilde{x}} g_0(s)K_0(s)ds \right] \ . \qquad (4.28)$$

Two independent solutions of the corresponding homogeneous equation can also be found in the same way. We may also find them by differentiation giving (for $k = 0$) an even and an odd solution

$$g_{1h}^{(1)} = \frac{dg_0}{d\tilde{x}} \left(= \frac{\partial g_0}{\partial \tilde{x}} \right) \ , \quad g_{1h}^{(2)} = \frac{\partial g_0}{\partial b} \ . \qquad (4.29a, b)$$

Note that the first solution is of dubious value (high-order singularity!) and the second meaningless when $g_0(\tilde{x} + k)$ is a solution with algebraic decay.

The case $g_0(\tilde{x}) = b \tanh \big(b(\tilde{x} + k)/2 \big)$ is discussed in Lagerstrom (1976). It was shown that for $\tilde{x}$ large, ϵg_1 contains a term $\epsilon \tilde{x} = x$ which matches with the outer solution. In fact, $g_0 + \epsilon g_1$ is a composite solution valid to order ϵ, whether $g_0(\tilde{x})$ represents a boundary layer or an interior layer. Assuming this we shall find $g_0 + \epsilon g_1$, with certain constants to be determined from the boundary conditions. The assumption about uniform validity will here be verified *a posteriori* by showing that the solution obtained contains the inner and outer expansions to the required order (see also Exercise 4.5).

Here two special cases[†] will be discussed: 1) The case of an interior layer, and 2) The case of a decreasing boundary layer at the left endpoint. In the second case

---

[†] The author received essential help with the analytical work from D. A. Reinelt (at the time a Caltech graduate student), and later from Caltech undergraduate Hanif S. Mamdani who also helped with the numerical work.

we shall be especially interested in very small values of b so that we are close to a boundary layer with algebraic decay.

The derivative with respect to b may be handled by the standard trick of just keeping $g_0(\tilde{x})$ but letting b have the form

$$b = b_0 + \epsilon b_1 , \quad b_0 \text{ and } b_1 \text{ independent constants} . \quad (4.30a)$$

Similarly the derivative with respect to $\tilde{x}$ is the same as the derivative with respect to the translational constant. However, we already have two constants of different order, namely $-x_d$ and k. Thus we put

$$x^* = \tilde{x} + k = \frac{x - x_d + \epsilon k}{\epsilon} , \quad x_d \text{ and } k \text{ independent.} \quad (4.30b)$$

Case 1. Interior Layer. We assume $-1 < A < 0$, $B = 1$. Evaluating (4.27) and (4.28) we obtain (writing g_1 instead of g_{1p} since the solutions of the homogeneous equation are obtained by the variation of the constants in g_0),

$$g_0 = b \tanh \frac{bx^*}{2} , \quad G_0 = \int_0^{x^*} g_0(s)ds = 2 \log \cosh \frac{bx^*}{2} ,$$

$$H_0 = \cosh^2 \frac{bx^*}{2} = \frac{\cosh bx^* + 1}{2} ,$$

$$K_0 = \frac{\sinh bx^*}{2b} + \frac{x^*}{2} = \frac{1}{b} \sinh \frac{bx^*}{2} \cosh \frac{bx^*}{2} + \frac{x^*}{2} , \quad (4.31)$$

and

$$g_1(x^*) = \left[\frac{2}{b} \tanh \frac{bx^*}{2} + x^* \operatorname{sech}^2 \frac{bx^*}{2} \right] \log \cosh \frac{bx^*}{2}$$

$$- \frac{1}{b} \tanh \frac{bx^*}{2} + \left[\frac{x^*}{2} - \frac{b}{2} \int_0^{x^*} s \tanh \frac{bs}{2} ds \right] \operatorname{sech}^2 \frac{bx^*}{2} . \quad (4.32)$$

The boundary values are chosen such that an interior layer occurs. Thus, at $x = 0$, $x^* = -x_d/\epsilon + k$ and at $x = 1$, $x^* = (1 - x_d)/\epsilon + k$. This gives the equations

$$A = -b_0 - x_d + \epsilon \left(-b_1 + \frac{2}{b_0} \log 2 + k + \frac{1}{b_0} \right) , \quad (4.33a)$$

$$B = b_0 + 1 - x_d + \epsilon \left(b_1 - \frac{2}{b_0} \log 2 + k - \frac{1}{b_0} \right) . \quad (4.33b)$$

The error is $o(\epsilon)$, and exponentially small terms e^{x^*} (x^* near $-\infty$) and e^{-x^*} (x^* near $+\infty$) have been neglected. The solution to (4.33) is

$$2x_d = 1 - A - B , \quad 2b_0 = -1 + B - A , \quad k = 0 , \quad b_0 b_1 = 2\log 2 + 1 . \quad (4.34)$$

Let x_c be the *crossing point* of the exact solution, that is the value of x for which u vanishes,

$$u(x_c; \epsilon) = 0 . \quad (4.35)$$

As ϵ tends to infinity (see Exercise 4.8) the exact solution tends to a straight line between the $x = 0$, $u = A$ and $x = 1$, $u = B$. In the case $A = -1$, $B = 1$, $x_d = 1/2 = x_c$ for any ϵ. However, if $A = -1/2$, $B = 1$, then $x_d = 1/4$. This is the value of x_c as ϵ tends to zero. When $\epsilon = \infty$, $x_c = 1/3$ and we expect x_c to move from $x_c = 1/4$ to $x_c = 1/3$ as ϵ increases. However, (4.34) shows that $k = 0$ so that including ϵg_1 shows no change in x_c.

Table 4.1 given below is instructive. It was found by numerical computation of the exact problem by Hanif Mamdani. (A similar table with different values of ϵ was computed by Caltech graduate student Elliot Fischer and discussed in Lagerstrom (1976). The value of x_c given there for $\epsilon = .5$ is between .3211 and .3212 .) See also Exercise 4.8.

ϵ	x_c	$(du/dx)_{x=0}$	$(du/dx)_{x=1}$	$(du/dx)_{x=x_c}$
0.1	0.28361	1.4557	1.0166	1.9602
0.05	0.26351	1.3186	1.0002	2.3644
0.01	0.25001	1.0010	1.0000	5.1408
0.005	0.25000	1.0000	1.0000	8.3979
0.001	0.25000	1.0000	1.0000	33.579
0.0005	0.25000	1.0000	1.0000	64.853

Table 4.1. Numerical computations for $A = u(0) = -1/2$, $B = u(1) = 1$.

Case 2. Decreasing Boundary Layer at Left Endpoint. We choose $A > 0$ and B such that the solution increases almost linearly in the right half of the interval studied (thus $0 < B \leq 1$). The calculations will be similar to those used for Case 1; however, the hyperbolic tangent will be replaced by the hyperbolic cotangent, and x_d will have the value zero. We shall be interested in the case for which the boundary is close to that of algebraic decay. This means that b will be small; how small it has to be will be determined by the calculations.

In terms of integrals we find

$$g(\tilde{x}) = b\coth\frac{b(\tilde{x}+k)}{2} + \epsilon\left[2\operatorname{csch}^2\frac{b(\tilde{x}+k)}{2}\int_k^{\tilde{x}+k}\sinh^2\frac{bs}{2}\log\sinh\frac{bs}{2}\,ds\right], \quad (4.36a)$$

or, integrating by parts,

$$\begin{aligned}
g(\tilde{x}) = {}& b\coth\frac{b(\tilde{x}+k)}{2} + \epsilon\left[\frac{2}{b}\coth\frac{b(\tilde{x}+k)}{2}\log\sinh\frac{b(\tilde{x}+k)}{2}\right.\\
& - (\tilde{x}+k)\operatorname{csch}^2\frac{b(\tilde{x}+k)}{2}\log\sinh\frac{b(\tilde{x}+k)}{2}\\
& - \frac{1}{b}(\sinh bk - bk)\operatorname{csch}^2\frac{b(\tilde{x}+k)}{2}\log\sinh\frac{bk}{2}\\
& - \frac{1}{b}\coth\frac{b(\tilde{x}+k)}{2} - \frac{(\tilde{x}+k)}{2}\operatorname{csch}^2\frac{b(\tilde{x}+k)}{2}\\
& + \frac{1}{2b}(\sinh bk + bk)\operatorname{csch}^2\frac{b(\tilde{x}+k)}{2}\\
& + \left.\frac{b}{2}\operatorname{csch}^2\frac{b(\tilde{x}+k)}{2}\int_k^{\tilde{x}+k}s\coth\frac{bs}{2}\,ds\right]. \quad (4.36b)
\end{aligned}$$

We know already from earlier calculations that the discontinuity occurs at $x = 0$ in the limit $\epsilon \downarrow 0$; this is why $x_d = 0$. We assume

$$b = o(1), \qquad k = O_s(1). \quad (4.37)$$

Evaluating (4.36a) for $\tilde{x} = 0$ we find, since b is small,

$$(g_0 + \epsilon g_1)_{x=0} = A = \frac{2}{k} + o(1). \quad (4.38)$$

Thus

$$k = \frac{2}{A}. \quad (4.39)$$

From (4.36b) we find at $x = 1$, $\tilde{x} = 1/\epsilon$,

$$B = b + \epsilon \left(\frac{1}{\epsilon} + k - \frac{1}{b} - \frac{2 \log 2}{b} \right) + o(\epsilon) \,.$$

Hence

$$b + \epsilon \left(k - \frac{1}{b} - \frac{2}{b} \log 2 \right) = B - 1 + o(\epsilon) \,,$$

$$b^2 = \epsilon(bk - 1 - 2 \log 2) \simeq b(B - 1) \,.$$

In the limiting case $B = 1$ we find

$$b^2 \simeq \epsilon(1 + 2 \log 2) \,. \tag{4.40}$$

Thus $b = O\left(\epsilon^{1/2}\right)$, which justifies dropping the term ϵbk above. If $B - 1 = O_s(1)$ then $b = O_s(1)$ and the assumption that b is small is not valid. Thus for the case $A > 0$, $B = 1$ the leading term of the inner expansion is a boundary layer with algebraic decay. The order of b is $\epsilon^{1/2}$, which still makes $b\tilde{x}$ a scaled variable relative to x. The outer solution is the straight line $f_0 = x$. The next approximation is given by (4.36). This formula was evaluated numerically by Hanif Mamdani for $A = 1$, $B = 1$, $\epsilon = 10^{-4}, 10^{-3}, 10^{-2}, 10^{-1}$. For $\epsilon = 10^{-4}$ he finds, to three decimals, that the boundary conditions are satisfied, $u = .020$ for $x = .010$, $u = .021$ for $x = .020$, $u = x$ for $x \geq .110$; u attains its minimal value $u_{\min} \doteq .02$ at $x = x_{\min} = .01$. Some other results are given in the table below (see also Exercise 4.9).

ϵ	$u(0)$	$x_{\min}$	$u_{\min}$	$u(1)$
10^{-3}	.999	.04	.06	1.002
10^{-2}	.996	.12	.19	1.020
10^{-1}	1.035	.26	.60	1.192

Table 4.2. Evaluation of Approximation (4.15) for various values of ϵ. $A = B = 1$.

Discussion of Special Features of Model Problem. The equation (4.1a) differs from the equations discussed in Section 2 by the nonlinear term uu'; if this term is replaced by $a(x)\,u'$ we are back to Section 2. As we have seen, the apparently harmless nonlinearity leads to a great variety of solutions.

First let us point out that the nonlinear term actually causes one simplification: the equation becomes autonomous. This leads to the possibility of reducing it to a first-order equation which actually is the constant of motion (4.23). This feature is, of course, not intrinsic in nonlinearity; other terms, say a forcing term, may be x-dependent. In an intuitive description of the example to be studied in Section 5 we consider a medium which is homogeneous (and isotropic). However, the assumption of spherical symmetry introduces the radial coordinates into the equation (while otherwise simplifying it).

In the linear case, if there is a boundary layer at one endpoint, say the left endpoint $x = 0$, we replace $a(x)\,u'$ by $a(0)\,u'$ in the boundary-layer equation. However, in our case, while $u(0) = A$ is known, it is essential to keep the term uu' intact. This also enables us to study how, with a suitable change of boundary values, the boundary layer detaches itself from the left endpoint and becomes an interior layer which finally attaches itself to the right endpoint. We have seen that our layer equation can describe a layer at an arbitrary point and also gives two-sided layers, a necessity for interior layers. These two essential properties would be lost if uu' were replaced by $u(0)\,u'$ in the inner equation.

The necessity for being able to deal with interior layers also forces us to consider the use of discontinuous outer solutions seriously. We can construct discontinuous outer solutions which have slope unity but which at a point x_d jump discontinuously from a value $f_0 = -a$ to $f_0 = b$. However, among those solutions, the only *relevant* ones are those for which $a > 0$ and $b = a$; these and only these match with a two-sided inner layer and are outer limits of exact solutions.

Relevant outer solutions may also be continuous but have discontinuous derivatives. These solutions have slope one, reach the value $f_0 = 0$ at a given point then remain zero in an interval of x, and may have a second discontinuity at which the solution changes back to the type whose derivative is unity. That such solutions are relevant can be seen from the Kaplun scaling (4.19); the importance of the scaling of the dependent as well as the independent variable is another typical nonlinear phenomenon. (In our case the scaling gave back the full equation, but this is not an essential feature.) We note that in this scaling $u = \epsilon^{1/2} h$. This does not imply that the outer limit u is zero: if h has a term $\bar{x}$ the outer limit of u is $x - x_d$.

Exercises for §4

4.1. Put $w = dg_0/d\tilde{x}$ and write (4.6) as a system of first order equations for g_0 and w. Find a first integral and sketch the corresponding curves in the (g_0, w)-plane. Put arrows in the direction of increasing $\tilde{x}$ (this is very instructive). Find all singular points. For each of the various types of solutions (4.7) find the corresponding class of integral curves.

4.2. Same as above, but use (4.5) with $\epsilon = 1/2$. Compare with Figure 4.1. Indicate how this phase-plane tends towards the phase-plane of (4.6) as $\epsilon \downarrow 0$. How do the curves change qualitatively? Which curves disappear? What happens to the singular points of Exercise 4.1?

4.3. Discuss the asymptotic solutions to leading order for the following lines in Figure 4.2.   a) $B = 1$ (first quadrant)   b) $B = -A + 1$ (fourth quadrant)   c) $B = -A - 1$ (fourth quadrant)   d) $B = A + 1$.

4.4. By a generalization of the method used in deriving Equation (1.23) justify Figure 4.3. Discuss the implications for matching of that figure.

4.5. Explain heuristically from a study of the pertinent equations why $g_0(\tilde{x}) + \epsilon g_1(\tilde{x})$ may be an approximation uniformly valid to order ϵ.

4.6. Write the basic equation (4.1a) in the variables h and $\bar{x}$. Since it is autonomous, apply the standard trick of regarding $\bar{x}$ as a function of h to reduce the order of the equation. Use the notation $p = d\bar{x}/dh$, $k = h^2/2$. The result is a simple case of Abel's differential equation which has a first integral equivalent to (4.23). Fill in the details. Repeat the same thing in terms of u and x to show how ϵ enters the results.

4.7. Let the invariant (4.23) play the role of a Hamiltonian. Derive the corresponding Hamilton's equations and compare with the original formulation of the problem.

4.8. Put $\eta = \epsilon^{-1}$. By a perturbation method find the first three terms in a solution of (4.1) which is uniformly valid as $\eta \downarrow 0$. Specialize to $A = -1/2$, $B = 1$ and discuss the crossing-point problem (Table 4.1) from the point of view of an expansion for ϵ large.

4.9. For $A > 0$, $B \geq 1$ give the leading-order composite expansion. Specialize to $A = B = 1$ and $\epsilon = 10^{-m}$, $m = 4, 3, 2, 1$. Evaluate the constant of integration C (see 4.23). Compare the results numerically with Table 4.2 and the discussion preceding that table. Indicate which curves in Figure 4.1 correspond to the boundary value problem here and how the curve varies with ϵ.

2.5. Model Equation for Flow at Low Reynolds Number

Introduction. The problem to be studied in the present section is

$$u'' + \frac{k}{x}\, u' + auu' + b(u')^2 = 0 ,\qquad (5.1a)$$

$$u(\epsilon;\epsilon) = 0 , \quad u(\infty;\epsilon) = 1 . \qquad (5.1b,c)$$

a, b, and k are constants with values to be discussed below. It was originally introduced by the author in an informal discussion group at Caltech in the 1950's as

a model[†] for flow at low Reynolds numbers; the particular values of the constants
were $a = 1$, $b = 0$, $k = 0, 1, 2$ (intuitively $k + 1$ represents the number
of dimensions in a spherically symmetric problem). The last term in (5.1a), with
$b = 1$, was added in a course given at Institut Henri Poincaré in Paris, written up in
Lecture Notes, Lagerstrom (1961). The object of adding the last term was to justify
a statement, made by Kaplun and Lagerstrom (1957), that "the Stokes equation is
linear only by accident." The problem has since been studied extensively; see for
instance Bush (1971), Hsiao (1973), Rosenblat and Shepherd (1975), Tam (1975),
MacGillivray (1978), the survey article by Lagerstrom and Casten (1972), and books
by Cole (1968) and Kevorkian and Cole (1981). It was also studied using rigorous
formal methods, letting k be any real number, by Cohen, Fokas and Lagerstrom
(1978); the phenomenon of expansion coefficients involving $\log \epsilon$ was discussed by
Lagerstrom and Reinelt (1984).

While (5.1) continues the sequence of problems modeling phenomena in fluid
dynamics, it also contains a further mathematical generalization of previous sections. Like the example in Section 4 the present example is nonlinear; even for
$b = 0$ it contains the quasilinear term uu' which is a typical transport term occurring in fluid dynamics. Unlike (4.1) it is nonautonomous (for $k \neq 0$); furthermore
the range of x is semi-infinite.

The small parameter ϵ does not occur in the equation, only in the position of
the left boundary. Another important version of the same problem is obtained by

---

[†] We remind the reader that "model" does not here mean a problem which yields
approximate quantitative results. Like Friedrichs' model in Section 1 and the model for
shock waves given in Section 4 it is meant to elucidate certain mathematical ideas used
in solving the associated physical problem. Friedrichs' example illustrated the matching
technique which Prandtl used in discussing problems of flow at high Reynolds numbers.
The present problem illustrates Kaplun's matching technique used in his solution of the
low Reynolds number problem.

introducing $\tilde{x}$ by

$$\epsilon\tilde{x} = x , \quad g(\tilde{x},\epsilon) = u(x,\epsilon) . \tag{5.2}$$

In the new variables the problem is

$$\frac{d^2g}{d\tilde{x}^2} + \frac{k}{\tilde{x}}\frac{dg}{d\tilde{x}} + \epsilon a g \frac{dg}{d\tilde{x}} + b\left(\frac{dg}{d\tilde{x}}\right)^2 = 0 , \tag{5.3a}$$

$$g(1,\epsilon) = 0 , \quad g(\infty,\epsilon) = 1 . \tag{5.3b, c}$$

The discussion of the original problem in fluid dynamics will be given in Chapter III Section 5. While we shall use here names from fluid dynamics, such as "Stokes equation," "Oseen equation" etc., the reader not familiar with fluid dynamics may think of these names as merely a distinctive terminology. However, it is important not to exclude the use of physical intuition. We therefore invent a physical problem to fit the equation. When k is a positive integer the first two terms of (5.1) represent the spherically symmetric part of the Laplace operator in $(k+1)$ dimensions. It is thus convenient to regard x as a radial variable and we choose to regard u as denoting temperature. The problem may then be viewed as a spherically symmetric problem of heat equilibrium with the last two terms representing nonlinear heat sources (or sinks). The temperature is kept at the value zero at the surface of the sphere $x = \epsilon$ (or $\tilde{x} = 1$) and has the value one at infinity. This interpretation is certainly fanciful, and it is doubtful whether any medium exists for which the temperature behaves as described; it still enables us to use valuable intuitive ideas. Of course, we have assumed all quantities to be nondimensional whereas in the original problem in fluid mechanics dimensional analysis gives important clues.

In our interpretation we have a medium which is being cooled by a sphere of radius ϵ; the resulting temperature distribution is stationary. Important for the understanding of the solutions to be described is the following heuristic principle: The larger the dimension $(k+1)$, the less the cooling power of the sphere. Examples will illustrate this: In one dimension $(k = 0)$ the surface of the sphere consists of the points $x = \pm\epsilon$, $(\tilde{x} = \pm 1)$. Imbedded in three dimensions, this surface corresponds to two parallel planes, separated by an x-distance of 2ϵ. In two dimensions the sphere is a circle which we think of as a circular cylinder in three dimensions; it obviously has a larger surface area than the corresponding three-dimensional sphere. A generalization of this principle to arbitrary integral dimension and even to real numbers is plausible. A (heuristic) corollary is that $(du/dx)_{x=\epsilon}$ (which is proportional to the heat transfer at a point on the sphere) increases with k. The reason is that a small area requires a large increase in the temperature near the sphere since u' must be large at $x = \epsilon$ in order for u to reach the value $u = 1$ at infinity. The heuristic ideas will later be verified in special cases.

We now discuss in analytical detail the various cases just mentioned, and then consider more general values of k. We notice that if $a > 0$, $b > 0$ their values may be rescaled to $a = b = 1$. However, $b = 0$ is an important special case. We shall therefore assume $a = 1$, $b = 0$ or $b = 1$; an important auxiliary equation will correspond to $a = 0$. Negative values of a and b will not be discussed.

Case of One Dimension $(k = 0)$: The sphere in one dimension, as imbedded in three dimensions, corresponds to two parallel planes which disconnect space into two unrelated halves and an "unphysical" region between the planes. Because of symmetry it is sufficient to consider $x \geq \epsilon$. The solutions of (5.1) are

$$b = 0 \; : \; u = \tanh \frac{x - \epsilon}{2}, \quad \left(\frac{du}{dx}\right)_{x=\epsilon} = \frac{1}{2}, \qquad (5.4a, b)$$

$$b = 1 \; : \; u = 1 - e^{-(x-\epsilon)}, \quad \left(\frac{du}{dx}\right)_{x=\epsilon} = 1. \qquad (5.4c, d)$$

We note that x and ϵ enter only in the combination $x - \epsilon$. It follows from our intuitive picture that the location of each plane is irrelevant. Mathematically, the explanation is that for the special value $k = 0$ the equation is autonomous, that is, invariant under translation in x. The parameter ϵ is irrelevant, and so is the statement that ϵ is small. Thus we are not dealing with a perturbation problem at all.[†]

Case of $k = integer \geq 1$: For dimensions ≥ 2 the problem is a genuine perturbation problem. It may be treated as a singular or as a regular perturbation problem, each method having certain distinctive advantages. The cases $b = 0$ and $b = 1$ will be discussed separately.

Historically (*i.e.*, in the original problem in fluid dynamics) the form corresponding to (5.3) for $b = 0$ was considered fundamental and a regular perturbation analysis was applied to this form. The leading term then obeys (5.3) with $\epsilon = 0$, an equation proposed by Stokes (1851). The equation, called the *Stokes equation*, was often considered to be derived by linearizing around the value of u at the inner boundary $(u = 0)$. This reasoning, repeated in textbooks for over a hundred years, depended on faulty or nonexistent dimensional analysis and will not be discussed in this section (see, however, Chapter III Section 5). Stokes found that in two dimensions $(k = 1)$ his equation had no solution; this is the so-called *Stokes paradox*. In three dimensions, however, Stokes found a solution which has been widely used and, as shown later, is actually a good approximation for small values of ϵ. However, in an attempt to find a higher-order approximation Whitehead (1889) found that the next term has a logarithmic singularity at infinity (*Whitehead paradox*). Oseen (1910) noticed the important fact that in the Stokes solution for $k = 2$ the terms neglected (corresponding to the term uu' in our model) actually are much larger for x large than the terms which are retained; in modern terminology the approximation used is nonuniform at infinity. Oseen therefore proposed to linearize around infinity which in our case (for $b = 0$) means replacing uu' by u' (see (5.5), (5.8), and (5.9)). The resulting equation, called the *Oseen equation*, has a solution for $k = 2$, given by Oseen, which actually improves the Stokes solution, and has also

---

[†] A similar situation will be encountered in Section 6 where we discuss the two-dimensional problem for the height of the meniscus.

a solution for $k = 1$ (Lamb, 1911). However, while Oseen's equation worked and his idea about the nonuniformity of the Stokes solution was correct, Oseen failed to give a systematic expansion procedure and did not explain why the Oseen equation is uniformly valid.[†] The conceptual structure of the problem was finally clarified by Kaplun whose basic ideas were first published in Lagerstrom-Cole (1955, "The Method of Kaplun" pp.873-874), and then announced at the Brussels 1956 ITAM meeting, see Kaplun (1957) and Kaplun-Lagerstrom (1957).

We now give an account of the application of Kaplun's ideas to (5.1). An essential point is that in any perturbation problem, including the singular ones, something is being perturbed and we ask what the unperturbed state is in our case. Kaplun's answer seems in retrospect simple: It is the medium without a body present.[‡] The temperature is then everywhere the same as the value given at infinity. The basic form of the equation is thus (5.1a), and the leading term of the outer expansion satisfies this equation and is

$$f_0 = 1 \, . \tag{5.5}$$

If now a very small sphere of radius ϵ with surface temperature $u = 0$ is inserted we expect (5.5) to be a good approximation except near the body on whose surface, $x = \epsilon$, u is prescribed to be equal to zero.[††] Thus the approximation (5.5) is not uniformly valid near the origin and we look for an inner variable to describe the field there. To find it we try scaling by $\eta(\epsilon)$, putting $\eta x_\eta = x$, $\eta \prec 1$. Our technique for

---

[†] His subsequent application of his equation to high Reynolds numbers was based on erroneous reasoning.

[‡] This was actually a major step. Oseen emphasized a nonuniformity of the Stokes solution at infinity. Kaplun emphasized the nonuniformity of (5.1) at $x = 0$; this led to a correct analysis.

[††] Conversely, as ϵ tends to zero the sphere, for $k \geq 1$, shrinks to a point and $u \to f_0$ for any fixed value of x. To put it graphically, *a point has no cooling power.*

finding distinguished equations introduced in Section 1 can be concentrated by the following reasoning: In all terms in (5.1) x appears as x^{-2} except in uu' which is of order x^{-1}. Thus for x_η fixed the third term is smaller than the others for any $\eta \prec 1$. This, however, does not pick out any specific η and we need some additional arguments. If $\eta \prec \epsilon$ the point x moves inside the body as $\epsilon \downarrow 0$ which is nonsensical. If $\eta = \epsilon$ the radius at the point x measured in radii of the body is constant; in particular at the surface of the body, where one of the boundary conditions is applied, x_η is always equal to unity. Intuitively, it is thus plausible that $\eta = \epsilon$ is a good choice and we choose as inner variable

$$\tilde{x} = x\epsilon^{-1} . \tag{5.6}$$

The full problem then takes the form (5.3) and g_0, the leading term of the inner approximation, satisfies the conditions

$$\frac{d^2 g_0}{d\tilde{x}^2} + \frac{k}{\tilde{x}} \frac{dg_0}{d\tilde{x}} + b \left(\frac{dg_0}{d\tilde{x}} \right)^2 = 0 , \tag{5.7a}$$

$$g_0(1) = 0 . \tag{5.7b}$$

This is the analogue of the *Stokes equation*. It is not linear for $b \neq 0$. On the other hand if the outer expansion is

$$u(x, \epsilon) = f_0 + \varsigma(\epsilon) f_1 + \cdots , \tag{5.8}$$

then f_1 satisfies what we call the (homogeneous) *Oseen equation* (for $a = 1$),

$$f_1'' + \frac{k}{x} f_1' + f_1' = 0 , \quad f_1(\infty) = 0 . \tag{5.9a, b}$$

We now summarize what the reasoning above has given us: The inner and outer variables are

$$\text{Outer variable is } x(\eta = 1) , \tag{5.10a}$$

$$\text{Inner variable is } \tilde{x} = x\epsilon^{-1} . \tag{5.10b}$$

The choice of outer variables was obtained by an intuitive reasoning which gave the unperturbed state. The same reasoning gave us f_0, the leading term of the outer expansion. The outer limit of the full equation gives us back the full equation. However, $f_0 = 1$ is a solution of that equation. If the beginning of the outer expansion has the form (5.8), then f_1 obeys the Oseen equation (5.9). The expansion coefficients, $\varsigma(\epsilon)$ and higher-order ones, will depend on k. A limit process keeping x_η fixed, $\eta \prec 1$, gives us the equation for the leading term of the inner expansion, namely the Stokes equation (5.7a). The inner variable $\tilde{x}$ is thus not found in the same way that distinguished variables were found in previous chapters; an intuitive plausibility argument was used to find (5.10b). *The Oseen equation is by nature linear, it is obtained by linearizing around the value at infinity; it is not by nature uniformly valid. The Stokes equation is not necessarily linear. It is in fact nonlinear for $b > 0$ (and linear "by accident" for $b = 0$).* The statements just made will be verified for $k = 1$ and 2. It is easy to verify directly from (5.4) that they are not applicable for $k = 0$. More general values of k will be discussed later on.

We shall first study the cases $k = 1$ and $k = 2$ in detail for $b = 0$.

Case $b = 0$, Two Dimensions $(k = 1)$: The Stokes equation is now

$$\frac{d^2 g_0}{d\tilde{x}^2} + \frac{1}{\tilde{x}} \frac{dg_0}{d\tilde{x}} = 0 , \tag{5.11}$$

whose solution, satisfying only the boundary at $\tilde{x} = 1$, is

$$g_0 = A \log \tilde{x} . \tag{5.12}$$

Obviously no choice of A can make (5.12) satisfy the outer condition at $\tilde{x} = \infty$ (the Stokes paradox). We therefore have to resort to matching with the outer solution $f_0 = 1$. At first sight matching also seems impossible since Prandtl's rule obviously fails, even in its generalization by Friedrichs. However, after having realized that this problem (or rather the original problem of the Stokes paradox and, more generally, the asymptotic problem of viscous flow at low Reynolds numbers) could be regarded as a singular-perturbation problem, Kaplun developed a powerful generalization of the Prandtl-Friedrichs technique, in the process introducing the extension theorem and concepts such as domain of validity, matching by overlap, intermediate variables, and many other related ones. While the application of these ideas may have seemed unnecessary and unduly sophisticated in previous problems, the present problem is suitable for concretely illustrating the meaning and the power of the ideas; an advantage is that here the analytical work is relatively simple compared to the original problem in fluid mechanics.

In the selection of an inner variable we found that the formal validity of the Stokes equation included any $\eta(\epsilon)$ such that $\eta \prec 1$ (although orders $\eta \prec \epsilon$ did not make intuitive sense). According to Kaplun's Ansatz (see Chapter I Section 4) this also represents the actual domain of validity. The domain of validity of the outer solution, which is obtained by applying the outer limit to the exact solution, includes $\eta = 1$. Thus by the extension theorem[†] (which is exact) there exist orders η_1 and η_2, $\eta_1 \prec 1 \prec \eta_2$, such that any η, $\eta_1 \prec \eta \prec \eta_2$, is in the domain of validity of f_0. We do not know a priori what the domain is; however, the existence of an extended domain is assured. Using Kaplun's Ansatz we conclude that there is an overlap domain in which both g_0 and f_0 are valid. We try to find a suitable η such that $\lim_\eta (f_0 - g_0) = 0$. We see that f_0 ($\equiv 1$) will cancel identically in the equation

$$f_0 - g_0 = 1 - A \log(x_\eta \eta \epsilon^{-1}) = 1 - A(\log x_\eta + \log \eta - \log \epsilon) , \qquad (5.13)$$

if we put

$$A = \phi(\epsilon) \overset{df}{=} -\frac{1}{\log \epsilon} . \qquad (5.14)$$

[†] Kaplun introduced the extension theorem precisely to handle the matching problem for the analogous problems for two-dimensional viscous flow.

The other terms in (5.13) must vanish because they are small. This condition is automatically fulfilled for $-A \log x_\eta$. We must now find an η such that $-\phi \log \eta = o(1)$. The choice $\eta = \epsilon^a$, $a > 0$, will not do, since the limit is $a > 0$. However, (see the last part of Chapter I Section 2) we can still fit in functions of ϵ which are $\succ \epsilon^a$, for all $a > 0$ but still $\prec 1$. One such function is $\phi(\epsilon)$, indeed, $\log |\log \epsilon| / |\log \epsilon| = o(1)$.

For higher-order terms we make the assumption that the outer and inner expansions are

$$\text{Outer}: \quad u \simeq 1 + \sum_{j=1} \phi(\epsilon)^j f_j(x) , \tag{5.15a}$$

$$\text{Inner}: \quad u \simeq \phi(\epsilon) \log \tilde{x} + \phi(\epsilon) \sum_{j=1} A_j \phi(\epsilon)^j \log \tilde{x} . \tag{5.15b}$$

We shall verify that this assumption is consistent for $j = 1, 2$ and 3. The outer expansion shows no arbitrary constants; they are simply absorbed in $f_j(x)$. We have assumed that all the $g_j(\tilde{x})$ of the inner expansion have the form $A_j \log \tilde{x}$, $A_j = $ constant. The reason is that in (5.3) g $(dg/d\tilde{x})$ is multiplied by ϵ which is transcendentally small compared to $\phi(\epsilon)$. Thus each $g_j(\tilde{x})$ will obey the same (homogeneous) Stokes equation, whose solution, since $g_j(1) = 0$, is $\log \tilde{x}$ times a constant. This does not mean that the nonlinear term uu' plays no role in deriving the inner expansion (5.15b); the constants A_j are obtained by matching with the outer expansion. For $j > 1$ the $f_j(x)$ obey nonhomogeneous Oseen equations in which the nonhomogeneous terms come from inserting the partial series $\sum_{l=1}^{j-1} \phi(\epsilon)^l f_l(x)$ into the term uu' of (5.1a). We note the possibility that to the expansions (5.15) one should add expansions involving coefficients ϵ^l which are transcendentally small relative to $\phi(\epsilon)$.

For $j = 1, 2$ we find from Appendix A5 and by matching

$$f_1(x) = -E(x) , \quad A_1 = \gamma \doteq .57721 , \tag{5.16a, b}$$

$$f_2(x) = -(1+\gamma)E(x) - e^{-x}E(x) + 2E(2x), \quad A_2 = \gamma^2 - 2\log 2 . \qquad (5.17a,b)$$

Further computations yield

$$f_3(x) = \left[(-\gamma^2 - 2\gamma + \frac{3}{2} + \log 2) + e^{-2x} - 2(\gamma + 1)e^{-x} \right] E(x)$$
$$- E(x)E(2x) - \frac{1}{2}e^{-x}(x+1)E^2(x) + 3\int_x^\infty \frac{e^{-2t}}{t} E(x)dt , \qquad (5.18a)$$

$$A_3 = \gamma^3 - \int_0^\infty e^{-t}E^2(t)dt + (1 - 6\log 2)\gamma + \frac{a}{2}\log 3 - 6\log 2 . \qquad (5.18b)$$

The value of the definite integral in (5.18b) is approximately 1.1254.

We postpone further discussion of the case $k = 1$ and proceed to the case $k = 2$, which corresponds to a spherically symmetric problem in three dimensions.

Case $k = 2$, $b = 0$; Switchback: As before $f_0 = 1$. The solution to (5.7) for the present case is $g_0 = A(1 - 1/\tilde{x})$. Putting $g_0(\infty) = 1$ gives $A = 1$. Since the inner solution need not be valid at infinity the assumption $g_0(\infty) = 1$ has to be justified; indeed matching $g_0(\tilde{x})$ with $f_0(x)$ also gives $A = 1$ for any $\lim_\eta$, $\epsilon \prec \eta \leq 1$. Thus the leading term of the inner expansion is

$$g_0(\tilde{x}) = 1 - \frac{1}{\tilde{x}} . \qquad (5.19a)$$

This solution corresponds to Stokes' classical solution for flow at low Reynolds numbers past a sphere.

By solving the Oseen equation we find that the beginning of the outer expansion is (see Appendix A5)

$$f_0 + \epsilon f_1 = 1 - \epsilon E_2(x) . \qquad (5.19b)$$

We make the natural assumption that the inner expansion starts $g_0(\tilde{x}) + \epsilon g_1(\tilde{x}) + O(\epsilon^2)$.

$$u \simeq g_0(\tilde{x}) + \epsilon g_1(\tilde{x}) + \cdots , \qquad (5.20)$$

Then g_1 satisfies the nonhomogeneous Stokes equation with an inner boundary condition,

$$\frac{d^2 g_1}{d\tilde{x}^2} + \frac{2}{\tilde{x}} \frac{dg_1}{d\tilde{x}} = -g_0 \frac{dg_0}{d\tilde{x}} \equiv \frac{1}{\tilde{x}^2}\left(\frac{1}{\tilde{x}} - 1\right) , \quad g_1(1) = 0 . \qquad (5.21)$$

The general solution with $g_1(1) = 0$ is

$$g_1(\tilde{x}) = -\frac{\log \tilde{x}}{\tilde{x}} - \log \tilde{x} + A_1 g_0(\tilde{x}) . \qquad (5.22)$$

Obviously, no choice of the constant A_1 can prevent g_1 from being logarithmically infinite at $\tilde{x} = \infty$.[†] However, as we have already seen for the case $k = 1$, the divergence of Stokes-type solutions at infinity does not invalidate our method provided matching can be carried out. In outer variables we find for small values of x

$$\epsilon^{-1}[(f_0 + \epsilon f_1) - (g_0 + \epsilon g_1)] = 1 - \gamma + O(x^2) + \frac{\epsilon \log x - \epsilon \log \epsilon}{x} + \log x - \log \epsilon - A_1\left(1 - \frac{\epsilon}{x}\right).$$

Thus we get overlap to order ϵ if we put $A_1 = 1 - \gamma - \log \epsilon$, and

$$\epsilon g_1(x) = -\epsilon(\log \epsilon) g_0(\tilde{x}) + \epsilon\left[(1 - \gamma)g_0 - \frac{\log \tilde{x}}{\tilde{x}} - \log \tilde{x}\right] . \qquad (5.23)$$

In a sense the assumption expressed by (5.20) is not wrong, but it is misleading if interpreted that the correction to $g_0(\tilde{x})$ is strictly of order ϵ. The determination of $g_1(\tilde{x})$ involves finding the constant of integration A_1 which multiplies the solution of the homogeneous equation. As we have already seen for the case $k = 1$ a constant of integration need not be strictly of order unity. It is a constant relative to the variable $\tilde{x}$ but may depend on ϵ. If we stick strictly to Poincaré-type expansions the correction to $g_0(\tilde{x})$, as shown by (5.23), has one term of order $\epsilon|\log \epsilon|$ and a term of the smaller order ϵ. However, the two orders come in simultaneously when matching is used to determine the missing constant of integration for the

---

[†] For the corresponding problem in fluid mechanics this is known as Whitehead's paradox.

problem (5.21). The term of order $\epsilon|\log\epsilon|$ has been called a "switchback term" by Kaplun. This is of course a subjective name: One expected the correction to be strictly of order ϵ but the guess was wrong; there is a larger correction of order $\epsilon|\log\epsilon|$. In fluid mechanics a similar term appeared when Proudman and Pearson (1957, p.255) used matched asymptotic expansions of the Kaplun type to study low Reynolds number (Re) flow past a circular cylinder: In addition to the expected term of order Re^2 they found a term of order $Re^2\log Re$; they realized clearly that this extra term was an automatic consequence of applying singular perturbation methods correctly. Van Dyke (1964, 1975) emphasized the prolification of similar logarithmic terms in many perturbation problems in fluid mechanics. Below we discuss the problem of switchback terms in a broader context.

Case $b = 0$, General Values of k: So far we have given k values which correspond to the familiar physical dimensions one, two and three. However, for a deeper understanding of the results it is important to consider the results for other values of k. In a paper by Cohen, Fokas and Lagerstrom (1978) k was allowed to be any real number (complex values were not considered) and b to be any number ≥ 0. The purpose of the study was to verify by rigorous mathematics certain results obtained by intuitive methods. We shall quote only one result here which will be of interest now. Above we used as an important tool in the calculations for $k = 1$ and 2 the assumptions that the outer limit of $u(x;\epsilon,k)$ is $f_0 \equiv 1$. This result is manifestly wrong for $k = 0$ and we may ask what happens for $0 < k < 1$. In the paper just mentioned it was proved that the result is true if and only if $k \geq 1$. The paper used the methods of rigorous non-intuitive mathematics: It did not provide any insight into the problem of why switchback terms occur or, more generally, why terms involving $\log\epsilon$ appear in certain expansion coefficients nor why the use of the Oseen equation to obtain uniformly valid results is allowed. We shall now discuss these problems without any claim of rigor.

Regular vs. Singular Perturbation Techniques. Switchback Terms. $k \geq 1$, $b = 0$. The restriction on k is made necessary by the remark made above. The restriction on b is not essential: By a transformation of the equation, to be discussed later, the reasoning below may in principle be applied; however, the analytical details are much more transparent if we assume $b = 0$. The discussion here is based mainly on that by Lagerstrom and Reinelt (1984). We first state

Proposition on Oseen Iteration. *Problem* (5.1), *for* $k \geq 1$ *and* $b = 0$, *may be solved as a regular perturbation problem using an Oseen iteration.* (5.24)

In discussing this we write (5.1a) as

$$L_k[v] \equiv \frac{d^2 v}{dx^2} + \frac{k}{x}\frac{dv}{dx} + \frac{dv}{dx} = -v\frac{dv}{dx} \ , \quad v = u - 1 \ . \tag{5.25}$$

By Oseen iteration we mean finding an expansion

$$v = 1 + v_1 + v_2 \cdots \ , \tag{5.26a}$$

by the iterative procedure

$$L_k[v_1] = 0 \ , \quad L_k[v_2] = -v_1' v_1 \ , \text{ etc.} \tag{5.26b, c}$$

The v_j are obviously closely related to the f_j; see Exercise 5.7. Equation (5.26b) is the Oseen equation; the boundary conditions are implied by the definition of v. Its solution, satisfying both boundary conditions, is

$$v_1 = \eta_1(\epsilon) E_k(x) \ . \tag{5.27}$$

The boundary condition at $x = \infty$ is satisfied automatically since the exponential integral (see Appendix A5) vanishes exponentially for x large. The boundary condition at $x = \epsilon$ need only be satisfied to a certain order in ϵ although if we put $\eta_1 = -1/E_k(\eta)$ it is satisfied exactly. We note that the inner limit of the Oseen equation gives the Stokes equation, that is *the Oseen equation contains (is richer than) the Stokes equation*. Thus using Kaplun's Ansatz (see Chapter I Section 4 and Chapter II Section 1), or rather an easy extension thereof, we conclude that the Oseen equation is uniformly valid to leading order and that it has a solution which is an approximation uniformly valid over the entire interval.[†] We also see that (5.26c) is the next outer equation; for $k > 1$ it also contains the inner equation (for $k = 1$ the inner equation would be homogeneous; we shall comment on this case briefly later). This shows the general idea of why the problem can be treated by a regular perturbation technique. While both the singular and the regular perturbation techniques are to be considered correct, the use of one of them may have practical advantages over the use of the other one depending on the specific question being studied. In the discussion to follow we shall find that alternate use of both techniques is convenient.

General Remarks on Logarithmic Switchback. We consider perturbation problems in which the small parameter ϵ, but not $\log \epsilon$, occurs in the formulation of the problem, whereas $\log \epsilon$ occurs in the expansion coefficients of some terms called logarithmic switchback terms. The expansion coefficients may have the form $|\log \epsilon|^{-k}$, $\epsilon^j |\log \epsilon|^k$, $\epsilon^j (\log |\log \epsilon|)^k$, etc., and need not occur multiplicatively (compare the discussion of the case $b = 1$ below). Such switchback terms have historically occurred conspicuously in the resolution of paradoxes in problems of fluid mechanics. Their origin has of course nothing to do in principle with fluid mechanics; a paradox is just a name for a result which is obviously wrong but has no apparent source of error, and its resolution often involves finding that a wrong expansion form has been assumed *a priori*; finally we have seen that even regular expansion methods

---

[†] While Oseen correctly pointed out a nonuniformity of Stokes solutions in the corresponding physical problem (Chapter III Section 5), he never faced the problem of why the equation he proposed could be uniformly valid or the problem of higher-order approximations.

may lead to log-terms in ϵ. We shall here make some general remarks to show that such terms are not unnatural. First we remark that the usual table of integrals of monomials x^k is misleading. The indefinite integral is another monomial except when $k = -1$ in which case $\log |x|$ results. This term seems like a surprising discontinuity among the monomials x^{1+k}. Let us instead consider the definite integrals,

$$z_k(x) = \int_1^x t^{-k}\, dt = \frac{x^{1-k} - 1}{1 - k}, \quad x > 0. \tag{5.28}$$

(We have used $-k$ instead of k to make the integral resemble that of the exponential integral). When $k = 1$ the right-hand side is indeterminate. However, as k approaches the value one (from above or below) the right-hand tends continuously to a definite function; we may take this as our definition of $\log x$ (we are only concerned with $x > 0$ and may hence omit the absolute value sign). For further details see Exercise 5.5.

To discuss the occurrence of $\log \epsilon$ in the expansion coefficients (for $b = 0$) we first use a regular perturbation technique, in particular the Oseen iteration, see (5.24)–(5.27). The leading term of a uniformly valid approximation to u is $1 + v_1 = 1 + \eta_1(\epsilon) E_k(x)$. It satisfies the boundary condition at infinity exactly; the inner condition is satisfied by choosing $\eta(\epsilon)$ such that $u(\epsilon)$ vanishes or at least is $o(1)$, which means that we may select the leading term in the expansion of $E_k(\epsilon)$ for ϵ small. In Appendix A5 the expansion for $E_k(x)$ for small values of ϵ is given. We see that it resembles that given by (5.28) and that the leading term, in fact any term, is logarithmic only when $k = 1$. From the nature of the Oseen expansion it seems obvious that powers of $(-\log \epsilon)^{-1}$ will appear in the higher-order expansion coefficients. A different approach to this phenomenon is given in Exercise 5.7, which relates the appearance of logarithmic switchback to the continuity argument used in discussing (5.28).

Considering now the case $k = 2$ we see that the present argument verifies what we have found earlier, namely that v_1 does not involve $\log \epsilon$. We know, however, that $\log \epsilon$ terms occur in higher-order expansion coefficients. From the present point of view this is due to the nonlinearity of (5.1a). While the equations for v_2 are linear the nonlinearity of (5.1a) gives rise to forcing terms which are products of E_k and E'_k. The response to these forcing terms involve functions $E_m(x)$, $m \neq k$, which when evaluated at $x = \epsilon$ give rise to $\log \epsilon$ terms. The situation is best studied with the aid of concrete examples (Exercises 5.8 and 5.9).

The examples just given indicate that logarithmic switchback terms occur in the Oseen iteration due to the fact that the expansion of $E_m(x)$ for small values of x has terms containing $\log x$ for integer values of m. Here m need not be the integer k occurring in the original equation; higher order iterations may contain terms such as $E_{2k-1}(x)$ where $m = 2k - 1$ is an integer (see Appendix). Various arguments indicate that the larger k is, the further the occurrence of a switchback term is postponed; the discussion above also suggests that switchback terms never occur when k is irrational. An estimate of when switchback occurs becomes unwieldy if we use the Oseen iteration (compare, for instance, (5.18)). However, in the previous discussion for $k = 2$ switchback terms were shown to be necessary by the occurrence of a $\log \tilde{x}$ term in (5.22), the expansion for $g_1(\tilde{x})$. This suggests the use of matched asymptotic expansions. We may study the inner expansion and can easily find out the first occurrence of a $\log \tilde{x}$ term even when certain constants, to be determined from the outer expansion, are not known. We may assume that the inner expansion has the form (5.20) for any $k > 1$ and that before the occurrence of a switchback term, subsequent expansion coefficients are powers of ϵ (Exercise 5.10). The equations for the successive terms are, for $k > 1$,

$$\frac{d}{d\tilde{x}} \left(\tilde{x}^k \frac{dg_j}{d\tilde{x}} \right) = -\tilde{x}^k r_j(\tilde{x}) , \qquad (5.29a)$$

$$r_0 = 0 , \qquad r_j = \sum_{m=0}^{j-1} g_m(\tilde{x}) \frac{dg_{j-1-m}}{d\tilde{x}} , \qquad \text{for } j > 0 . \qquad (5.29b)$$

The integration amounts simply to integrating monomials. The resulting terms may diverge for large $\tilde{x}$; this is allowed since we know that the condition at $\tilde{x} = \infty$ is replaced by a matching condition. However, a logarithmic divergence, involving $\log \tilde{x} = \log x - \log \epsilon$ necessitates the occurrence of $\log x$ in the outer expansion, evaluated for small values of x. Such a term at $x = \epsilon$ leads to $\log \epsilon$ terms in the expansion coefficients.

The discussion of switchback terms given above is based mainly on Lagerstrom-Reinelt (1984), to which we refer the reader for further details.

Positive Values of b $(b = 1)$. The reason for adding the term $b\left(u'\right)^2$ was discussed in the Introduction. For $b > 0$ it is sufficient to consider $b = 1$. The resulting Stokes equation and the inner boundary condition are given by (5.7). As observed in Cohen, Fokas and Lagerstrom (1978) the nonlinear Stokes equation is transformed into the linear one (for $b = 1$) by the change of the dependent variable,

$$g_0 \to e^{g_0} . \tag{5.30}$$

The solutions of $(5.7a, b)$ are then found to be

$$g_0 = \log \left(B \log \tilde{x} + 1 \right) , \qquad \text{for } k = 1 , \tag{5.31a}$$

$$g_0 = \log \frac{B\tilde{x}^{1-k} + 1 - k - B}{1 - k} , \qquad \text{for } k \neq 1 , \tag{5.31b}$$

where B is a constant of integration actually representing the value of $dg_0/d\tilde{x}$ at $\tilde{x} = 1$. (We note in passing that (5.31a) follows from (5.31b) as is seen by employing the limit process $k \to 1$, mentioned in the discussion of switchback.)

Since the homogeneous Oseen equation remains linear, it is easily solved; the first nonhomogeneous Oseen equation is different from the case $b = 0$, but its solution can be found from the formulas given in the Appendix. Much of the discussion for $b = 0$ can now be easily applied to the case $b = 1$ (see Exercises). We give some further comments below.

Comments on Poincaré-Type Expansions, Composite Expansions, Limit Processes.
The solutions (5.31) of the Stokes equations for $b = 1$ show that B, the constant of
integration, is not multiplicative; this is of course a consequence of the nonlinearity
of the equation. Thus if B depends on ϵ, which is the case for $k = 1$, the expansion
coefficient is not multiplicative and hence the inner expansion is not of the Poincaré
type. A more general type of asymptotic expansion thus occurs naturally and, as
we shall see, this may be advantageous from a practical point of view. Let us first
recall what happens for $b = 1$ and $b = 0$. A proposed inner expansion for the
case $b = 0$ is given by (5.15b), whose form reflects the fact that all terms obey the
homogeneous Stokes equation. The expansion (5.15b) may be written

$$u \simeq \Psi(\epsilon) \log \tilde{x}, \qquad \Psi(\epsilon) \simeq \sum_{j=0} A_j \phi(\epsilon)^{j+1}. \qquad (5.32)$$

Here $\Psi(\epsilon)$ is the value of $dg/d\tilde{x}$ at $\tilde{x} = 1$, a quantity which may be physically
important (in our interpretation it is proportional to the heat-transfer at the surface
of the sphere $x = \epsilon$ (or $\tilde{x} = 1$)). Of course, $\Psi(\epsilon)$ is not found in computation;
the coefficients A_j have to be evaluated, one by one, by matching with the outer
expansion, which, however, leads to an asymptotic approximation of the derivative.
The numerical accuracy of the result may be improved by recasting the partial series
into a different form. In his solution of the Stokes paradox, Kaplun (1957) computed
the second approximation and replaced the $\Phi(\epsilon) + A_1 \Phi(\epsilon)$ where $\Phi(\epsilon) = -1/\log \epsilon$,
by $\tilde{\Phi}(\epsilon) = 1/(-\log \epsilon + C)$, $C =$constant, which is asymptotically equivalent but
which gives better agreement with the experimental data. This process is discussed
in some detail by Van Dyke (1975, pp 243–247), who calls it telescoping.

Returning now to the case $b = 1$ we observe that, for $k = 1$, terms of the
expansion obey linear equations, except the first one. The result is much more
transparent if we rewrite the inner expansion in the form (5.31a) where now, however, $B(\epsilon)$ is a function of ϵ which can be determined to successively higher order in

ϵ, as was the case with $\Psi(\epsilon)$ in (5.32). The reason is the same as for $b = 0$: The term $\epsilon g\,(dg/d\tilde{x})$ is transcendentally small relative to $\Phi(\epsilon) = -1/\log \epsilon$. Thus, even if higher order terms in the inner expansion obey linear equations it is convenient, for $k = 1$, to rewrite any partial expansion in the concise form (5.31a). After this has been done, one may apply telescoping to any partial series of $B(\epsilon)$.

In the context of this book, however, the advantage of putting the result in a concise form whose physical meaning is transparent should be stressed.[†] *Systematic ideas for judging or improving numerical accuracy have not been introduced here.*

Above we stressed that, in our problem, an asymptotic expansion not of the Poincarè type occurs naturally and may have certain advantages. However, it may be of interest, and important in other problems, to investigate how an equivalent expansion of the Poincarè type may be constructed. This problem was solved by Bush (1971) who introduced the following transformation,

$$t = -\log x\,, \qquad u(x;\epsilon) = v(t;\epsilon)\,. \tag{5.33}$$

Then (5.1), for $a = b = k = 1$ becomes

$$\frac{d^2 v}{dt^2} + \left(\frac{dv}{dt}\right)^2 - \exp(-t)v\frac{dv}{dt} = 0\,, \tag{5.34a}$$

$$v(-\log \epsilon;\epsilon) = 0\,, \qquad v(-\infty;\epsilon) = 1 \tag{5.34b, c}$$

The inner expansion then takes on the Poincarè form naturally; Bush's transformation is also discussed in Lagerstrom-Casten (1972, p106ff.). We shall not try to evaluate here the merits of such a transformation; for a given problem this depends very much on what questions are being asked. It is, however, important to realize that the general ideas of matched asymptotic expansions, as presented in this book, are not necessarily tied to the use of Poincarè-type expansions.

[†] See the corresponding conjecture in Cohen, Fokas and Lagerstrom (1978, p206).

In Section 1 we introduced the concept of composite expansions, which are uniformly valid in the entire domain considered in the original problem. The name "composite" was used earlier because they were formed by combining inner and outer expansions in such a way that terms which represented a duplication were omitted. One way of doing this was to consider correction layers rather than the full boundary layer, or more generally to construct an inner (partial) expansion of the full function minus a corresponding partial outer expansion. In the discussion of the "Second Example" in Section 3 we stated that the usefulness of correction layers varies considerably. In Section 1 they turned out to be practical and instructive; however, in the present section, forming an inner expansion of $u - (f_0 + \varsigma(\epsilon)f_1 + \cdots)$ is utterly useless for finding a uniformly valid expansion. However, it turned out that for $b = 0$ the Stokes equation is contained in the Oseen equation, the outer expansion contains the inner expansion and is hence uniformly valid. Thus actually the whole inner expansion duplicates information already contained in the outer expansion. This is somewhat of a freakish situation.[†] For $b = 1$ none of the previously used methods seem suitable to combine a uniformly valued composite expansion. However, we observed earlier that in the case $b = 1$ the transformation (5.30) reduced the nonlinear Stokes equation to a linear one in a new dependent variable. Skinner (1981) noticed that if the same transformation is applied to the whole equation (5.1a) one obtains an equation whose corresponding Stokes equation is linear and contained in the Oseen equation. Like Bush's transformation in (5.33) this is a clever trick; while such methods seem of limited applicability, the two transformations should be analyzed further.

Some Equivalent and Related Equations. In the previous section we found that a

---

[†] An even more special case occurred in Section 4; it was seen there that for a large class of boundary conditions two terms of the inner expansion contained the whole outer expansion.

special scaling, which depended on the nonlinearity of the equation (the Kaplun scaling) played an essential role. A natural question is whether the nonlinearity of the equation of the present problem leads to an analogous scaling. The answer is that a new scaling does not seem to be necessary, at least not for the special boundary-value problem studied here; we observe also that the importance of the Kaplun scaling depended on the fact that the outer equation of the previous section had special solutions of the type given by Equation (4.13).

However, the same arithmetic used for studying the possible scalings in the previous section shows that the present equation, *for the special case* $b = 0$ is invariant under a similarity transformation which leaves the product ux invariant. The equation may then be reduced to a first-order system. Specifically, if we introduce the new variables,

$$y = xu, \quad v = x^2 \frac{du}{dx}, \tag{5.35}$$

then (5.1a), with $a = 1$ for convenience and $b = 0$ by necessity, reduces to

$$\frac{dy}{dv} = \frac{v + y}{v(2 - k - y)}. \tag{5.36}$$

A second change of variables transforms it into an Abel differential equation,

$$\frac{dw}{dy} + (y + k - 3)w^2 + y(y + k - 2)w^3 = 0, \tag{5.37a}$$

where

$$w = -\frac{1}{v + y}. \tag{5.37b}$$

Finally, the equation can be transformed into an equation which describes the equilibrium temperature distribution in a spherical gas cloud (star) under self-induced gravitational attraction, studied by Lemke (1913).[†]

$$\frac{d^2t}{dz^2} + \left(1 + \frac{2}{k-2}\right)\frac{1}{z}\frac{dt}{dz} + \frac{4}{(k-2)^2}e^t = 0, \qquad (5.38a)$$

where

$$y = \frac{(k-2)z}{2}\frac{dt}{dz}, \quad v = z^2 e^t. \qquad (5.38b, c)$$

Summary of Special Features of Model Problem. Compared to the nonlinear problem of the preceding section, the problem discussed here has certain new features described in the second paragraph of the present section. However, the detailed study of the problem demonstrates how that description misses out on some fundamental points. First, we must emphasize that the essential nature of the problem is best seen in the variables used in (5.1). The formulation in (5.3) is less transparent though equivalent. (As pointed out earlier, use of the second formulation obstructed for many years the solution of the corresponding problem in the theory of viscous flow.) The left endpoint, $x = \epsilon$, should not be viewed as just a point which happens to vary with ϵ. We choose to regard it as a sphere in $(k+1)$ dimensions which vanishes as $\epsilon \downarrow 0$. This geometrical point of view makes direct sense only when k is a positive integer. However, a more objective but less intuitive description is that as $\epsilon \downarrow 0$ the left endpoint tends to zero and $x = 0$ is a singular point of the equation. That $k = 0$ is a distinguished value of k is easy to see. It is less obvious that $k = 1$ is a distinguished value since the problem is singular for $k \geq 1$ and nonsingular, as defined in the text, for $k < 1$.

This question, as well as that of logarithmic switchback, is connected with the question of the behavior of the solution for x small. This is a standard topic in

---

[†] While the author was on a leave of absence, the problem (5.1) with $a = 1$ and $b = 0$ was discussed by J. Kevorkian, D. MacGillivray and Y. M. Lynn. The transformations introduced above, as well as the reference from Lemke, are taken from unpublished notes by Kevorkian.

the study of second-order linear ordinary differential equations. The theory there tells us whether there is a basic solution which is logarithmic at $x = 0$. For the nonlinear case the general solution is no longer a linear combination of two basic solutions and, in our case, the occurrence of logarithmic terms in higher order terms of the expansion near zero depends on the nonlinearity of the equation.

In studying logarithmic switchback we found that it is advantageous to use two different viewpoints, using both a regular and a singular perturbation technique. This is also of importance for another vexing problem, namely how to combine inner and outer expansions to obtain a uniformly valid composite expansion. This cannot be done by any of the methods used in previous sections. Our only answer was to find an outer expansion which *is* composite, that is, contains the inner expansion. This is straightforward for $b = 0$ but for $b > 0$ a trick, based on a linearizing transformation of the inner equation, was needed.

Regarding the inner expansion as a boundary-layer type expansion may be inappropriate terminology: Even in our multidimensional interpretation, the layer does not occur near a boundary, but near a singular point. In the next two sections we shall deal with other examples of inner solutions near singular points; the first example will show certain obvious similarities to the present one, the second example will be radically different. The name "boundary-layer" still seems a convenient term for the inner layer in all examples.

It was pointed out in the last paragraph of Chapter I Section 3 that having the small parameter ϵ multiplying the highest derivative of the differential equation is neither a necessary nor a sufficient condition for a problem to be singular (actually we were speaking specifically about problems for which the use of layer-type techniques was natural). The present problem has two formulations: (5.1) where ϵ occurs only in the boundary condition, and (5.2) where ϵ occurs only as a coefficient of a lower derivative. We saw that the problem may be treated by a regular perturbation technique or by a singular perturbation technique. The technique of matched asymptotic expansion used here is that formulated by Kaplun; while the ideas needed for Section 1 are simpler than the ones needed here, both apply special cases of some very general ideas.

Finally we note that while switchback phenomena first became prominent in singular-perturbation problems our example shows that the same phenomena may occur when a regular perturbation method is used.

A5. Appendix

The exponential integral $E_k(x)$ is defined by

$$E_k(x) = \int_x^\infty t^{-k} e^{-t} dt \,. \tag{A5.1}$$

$E_1(x)$ is usually written simply as $E(x)$. For $k > 1$, $k \neq$ integer, the integral may be expanded as

$$E_k(x) = C_k + \sum_{j=1}^\infty \frac{(-1)\, x^{j-k}}{(j-1)!\,(j-k)} \,, \tag{A5.2a}$$

where

$$C_k = \int_0^\infty t^{-k} \left[e^{-t} - \sum_{j=0}^{[k]-1} \frac{(-t)^j}{j!} \right] dt \,, \tag{A5.2b}$$

$$[k] = \text{greatest integer} < k \,. \tag{A5.2c}$$

The inhomogeneous Oseen equation

$$\frac{d^2 f}{dx^2} + \frac{k}{x}\frac{df}{dx} + \frac{df}{dx} + E_k \frac{dE_k}{dx} = 0 \tag{A5.3}$$

has the particular solution with $f(\infty) = 0$

$$f = F_k(x) \equiv \frac{dE_{k-1}}{dx} E_k(x) + 2^{2k-1} E_{2k-1}(2x) \,, \tag{A5.4}$$

and the equation

$$\frac{d^2f}{dx^2} + \frac{k}{x}\frac{df}{dx} + \frac{df}{dx} + \left(\frac{dE_k}{dx}\right)^2 = 0 \qquad (A5.5)$$

has the solution

$$f = G_k(x) = -\frac{1}{2}(E_k(x))^2 . \qquad (A5.6)$$

Asymptotic expansions valid for x small are

$$E_1(x) = -\log x - \gamma + x + O\left(x^2\right) , \quad \gamma = \text{Euler's constant} \doteq 0.5772 , \quad (A5.7a)$$

$$E_2(x) = \frac{e^{-x}}{x} - E_1(x) = \frac{1}{x} + \log x + (\gamma - 1) - \frac{x}{2} + O\left(x^3\right) , \qquad (A5.7b)$$

$$E_3(x) = \frac{e^{-x}}{2x^2} - \frac{e^{-x}}{2x} + \frac{1}{2}E(x) = \frac{1}{2x^3} - \frac{1}{x} - \frac{1}{2}\log x + \frac{3-2\gamma}{4} + \frac{x}{6} + O\left(x^2\right) . \quad (A5.7c)$$

Exercises for §5

5.1. Verify that $f_0 - g_0$ in (5.13) tends to zero if $\eta = 1$ (outer limit) even though this function is outside the overlap domain. Discuss this in the light of the last paragraphs of Chapter I Section 2.

5.2. Verify (5.16), (5.17), and (5.18).

5.3. Find the partial inner expansion contained in the partial outer expansion as given by (5.16)–(5.18).

5.4. For $k = 3$, $b = 0$ solve the Stokes equation and its iterations satisfying both boundary conditions as far as possible. Find the lowest value of j such that $g_j(\tilde{x})$ cannot be made to satisfy the condition at $\tilde{x} = \infty$. Give a heuristic explanation why j increases with k. Discuss the corresponding outer expansion and resolve the apparent paradox that for the critical j, $g_j(\tilde{x})$ cannot satisfy the outer boundary condition.

5.5. a) Investigate the uniformity in x of the convergence $z_k(x) \to \log x$ as $k \to 1$ (see Equation 5.28).

b) Compute $x_k(z)$, the inverse of $z_k(x)$. Find the limit of $x_k(z)$ as $k \to 1$.

5.6. Discuss the relations between the v_j of the Oseen iteration and the f_j of the outer expansion.

5.7. For $k = 1$ find f_1 and f_2 by a limit process from the solutions for k slightly greater than one.

5.8. The logarithmic switchback term occurring for $k = 2$ was studied with the aid of a singular perturbation technique. Derive the same result by using the Oseen iteration for values of $k \neq 2$ but near 2 and letting $k \to 2$.

5.9. Find a switchback term for $k = 3/2$ using both methods employed here for the case $k = 2$.

5.10. Show that for $k > 1$ the partial inner expansion has the form $\sum\limits_{j=0}^{} \epsilon^j g_j(\tilde{x})$ before switchback occurs.

5.11. We consider (5.1) with $a = 1$, $b = 1$.   a) Get three terms of the outer expansion for $k = 1$ and the corresponding terms of the inner expansion. b) Same problem for $k = 2$. Comment on switchback.

2.6. Stationary Meniscus Inside and Outside of a Tube

Introduction. In this section we discuss the problem of the shape of the stationary surface of a liquid in a very large container ("reservoir") into which a circular cylindrical tube has been inserted vertically. The governing forces are hydrostatic pressure determined by gravity, surface tension, and adherence of the liquid to the inside and the outside solid walls of the tube. Thus we study a genuine physical problem: It is a mathematically formulated problem whose solutions give an approximate quantitative description of experimentally observable facts. In the context of this book we are mainly concerned with values of the Bond number, the nondimensional parameter to be defined below, for which the use of perturbation methods is reasonable. We shall thus consider four cases: the meniscus inside and outside the tube at large and small Bond numbers. Actually, this will provide an interesting variety of problems for illustrating the use of regular and singular perturbation techniques; occasionally we shall meet again ideas and techniques which played a significant role in the preceding section (and to which we shall return in the discussion of viscous flow at low Reynolds numbers in Chapter III, Section 5), and various new ideas will also be encountered. The interior problem for large Bond numbers is of special historical interest: In his pioneering investigation of surface tension Laplace studied that problem by what seems to be the first successful use of matched asymptotic expansions.[†]

The specific problem to be discussed is the following: A circular[‡] tube is inserted vertically into a liquid in a very large container; the liquid is in contact with a gas. We study the asymptotic theory of the shape of the meniscus, that is, of the surface of the liquid separating the liquid from the gas. Only the *stationary* state is considered (no capillary waves). The shape is determined by the pressure difference between the liquid and the gas, by the surface tension at the liquid surface, and by the adherence of the liquid to the solid walls of the tube. We make the following assumptions: The *density of the liquid is a constant ρ*; the pressure of the liquid is hydrostatic, the variations in the pressure of the gas are negligible; the region separating the liquid and the gas is treated as a mathematical surface; *the surface tension coefficient, σ, is a material constant of the liquid*; the extent of the container is considered infinite, that is, the influence of its walls may be neglected and at sufficiently large distances the influence of the solid tube vanishes; the tube is

---

[†] Apparently, Laplace did not think of his analysis as introducing a new method, capable of various applications. Compare the comment on Prandtl's boundary-layer theory in the historical remarks in Chapter III, Section 5.

[‡] We shall also briefly treat the corresponding two-dimensional problem in which the tube consists of two parallel flat plates.

open at upper and lower ends, in particular the liquid inside the tube is in pressure contact with that outside the tube.

We combine various constant parameters to form a parameter of dimension length, called the *capillary length* l_c, and defined by

$$l_c^2 = \frac{\sigma}{\rho g} \, , \qquad\qquad (6.1a)$$

where $g = acceleration\ of\ gravity$ (positive downward). The inserted tube has a characteristic length which we call the *geometrical length*, l_g. We choose it as

$$l_g = \text{radius of inserted tube}\,. \qquad\qquad (6.1b)$$

From the two lengths we construct the dimensionless parameter,

$$B = \frac{l_g^2}{l_c^2} = \frac{\rho g l_g^2}{\sigma} = \text{Bond number}\,; \qquad\qquad (6.2a)$$

occasionally we use instead

$$\epsilon = B^{-1/2} = \frac{l_c}{l_g}\,. \qquad\qquad (6.2b)$$

B, or ϵ, will be our principal parameter: We shall consider asymptotic solutions for $B \prec 1$ and for $\epsilon \prec 1$. A second dimensionless parameter is also important; it is considered a given material constant,

$$\theta = \text{contact angle (between meniscus and tube)}\,. \qquad\qquad (6.2c)$$

We shall assume $0 \le \theta < \pi/2$ (*wetting liquid*) which is the case shown in Figure 6.1. The case of a *non-wetting liquid* (such as mercury), $\pi/2 < \theta \le \pi$, can be treated by reduction to a case of a wetting liquid, see Concus (1968); $\theta = \pi/2$ is of course trivial.

The coordinates to be used are shown in Figure 6.1; subscript "d" denotes "dimensional," $\theta = \frac{\pi}{2} - \psi_w$ where subscript "w" denotes value at the wall of the

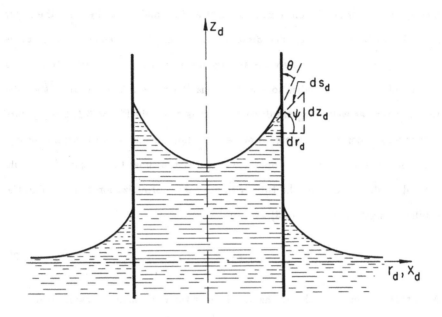

Figure 6.1. Dimensional Coordinates for Meniscus. Wetting Liquid.

tube, s is the length measured along the intersection of the meniscus and a vertical plane through the axis of the tube.

Physical Background. The foundations for a physical theory of the meniscus were laid by Laplace and by Young (for historical details see Finn (1986)). The discussion below is based on the concise treatment by Landau and Lifshitz (1959).

The basic formula, called Laplace's formula (or, occasionally, the Laplace-Young formula) is

$$\sigma(\kappa_1 + \kappa_2) = \text{pressure difference} = -\rho g z_d, \qquad (6.3)$$

which is valid at the meniscus. Here κ_1 and κ_2 are the two principal curvatures of the meniscus (the sign convention used will be obvious in each special case). The pressure difference is the difference between the pressure of the gas and that of the liquid; the former is considered constant and normalized to be zero, the latter is the hydrostatic pressure. The *dimensional height of the meniscus, called* z_d, is normalized in (6.3) so that its value at large distances from the inserted tube tends to zero; alternative normalizations will be considered later. At the walls of the tube the curvature as well as z_d are discontinuous as shown in Figure 6.1; the pressure is then also discontinuous there. However, the liquid inside the tube communicates with the liquid outside at the open lower end of the tube; thus Laplace's formula is valid inside and outside the tube with the same normalization of z_d. Using the capillary length we rewrite (6.3) as

$$\kappa_1 + \kappa_2 + \frac{z_d}{l_c^2} = 0 \,. \tag{6.3'}$$

Note that the terms on the left-hand side have the dimension inverse length.

Further notation and terminology. The dimensional horizontal coordinate is denoted by r_d and x_d for the case of cylindrical symmetry and the two-dimensional case, respectively. Since we shall use singular-perturbation techniques we anticipate that both length scales, l_c and l_g, will be needed in the definitions of nondimensional quantities. Thus we introduce two types of independent variables,

$$x^* = \frac{x_d}{l_c} \,, \quad r^* = \frac{r_d}{l_c} \,, \tag{6.4a}$$

and

$$x = \frac{x_d}{l_g} = \epsilon x^* \,, \quad r = \frac{r_d}{l_g} = \epsilon r^* \,. \tag{6.4b}$$

The vertical coordinate z_d is a dependent variable. To relate it to the contact angle we need its value at the wall, that is at $|x_d| = l_g$. A natural nondimensional variable is therefore z as defined below (but z^* will also be used),

$$z^* = \frac{z}{l_c} \,, \quad z = \frac{z_d}{l_g} = \epsilon z^* \,. \tag{6.4c, d}$$

The use of nondimensional notation is convenient for writing concise formulas; however, the use of dimensional concepts occasionally facilitates intuitive reasoning. The two methods may be combined: We may refer, say, to the case of very small Bond number as the case of a very narrow tube. For instance, we may consider l_c fixed while l_g tends to zero. Alternatively we may think of the case of small B as the case when g tends to zero while the other dimensional parameters remain fixed. Examples of both interpretations will be shown below.

Two-dimensional Case. The inserted tube now consists of two parallel vertical planes. This case is of direct physical interest only as a limiting case. It is analytically simpler than the case of cylindrical symmetry (circular tube), and it is conceptually and analytically instructive to study it.

We denote the dimensional horizontal coordinate by x_d rather than by r_d; the point $x_d = 0$ is located halfway between the two plates and the geometric length l_g is the value of $|x_d|$ at either plate. In Laplace's formula one principal curvature is the curvature of the graph z_d versus x_d and the other principal curvature is zero. Thus in the special case considered here (6.3) and (6.3') become

$$-\kappa_1 \equiv \frac{z_d''}{\left(1 + (z_d')^2\right)^{3/2}} = \frac{z_d}{l_c^2} , \tag{6.5}$$

where "prime" denotes "$\frac{d}{dx_d}$."

We follow essentially the discussion by Landau and Lifshitz but use also an idea from Rayleigh (1915): Both z_d, x_d and the arc length s_d will be expressed as functions of the angle ψ, see Figure 6.1.[†] Thus instead of (6.5) we write

$$\frac{z_d}{l_c^2} = -\kappa_1 = \frac{d\psi}{ds_d} = \frac{d\psi}{dz_d} \sin \psi = -\frac{d\cos \psi}{dz_d} , \tag{6.6a}$$

$$\frac{z_d}{l_c^2} = \frac{d\psi}{ds_d} = \frac{d\psi}{dx_d} \cos \psi . \tag{6.6b}$$

[†] Landau and Lifshitz introduce ψ in one case as a formal trick in changing the variable of integration while here it has a clear geometrical interpretation.

The first equation gives

$$\frac{z_d^2}{2} = l_c^2(A - \cos \psi), \quad A = \text{constant of integration}. \tag{6.7a}$$

Then the second equation gives

$$x_d = l_c 2^{-1/2} \int_0^{\psi} \frac{\cos \phi \, d\phi}{(A - \cos \phi)^{1/2}}. \tag{6.7b}$$

The integral may be evaluated in terms of elliptic functions. From this we may relate the constant of integration A to the radius l_g and the contact angle θ by using the boundary condition

$$x_d = l_g \text{ at } \psi = \psi_w = \frac{\pi}{2} - \theta. \tag{6.7c}$$

Thus the problem is solved in principle. However, in the present context our main aim is to obtain relatively simple approximate formulas asymptotically valid for very small or very large values of the Bond number B. We now discuss this for the *interior* and the *exterior* problem separately. It will be seen that the latter problem has certain peculiar features, tied specifically to the fact that we consider a two-dimensional problem. Similar features were observed in the problems discussed in Section 5 for the special parameter value $k = 0$.

We start with the *interior* problem. Using the solution implicitly contained in (6.7) we shall relate A, the constant of integration, to the principal parameter B. First consider the case $B \prec 1$ (narrow tube). From (6.7b) we see that l_g very small implies that A is very large. Thus to leading order in B we may write (6.7b) as

$$\sqrt{2}\, x_d \sim l_c A^{-1/2} \int_0^{\psi} \cos \phi d\phi = l_c A^{-1/2} \sin \psi. \tag{6.8a}$$

The boundary condition at the tube gives

$$\frac{\sqrt{2}\, l_g}{l_c} = \sqrt{2B} \sim A^{-1/2} \sin\left(\frac{\pi}{2} - \theta\right) = A^{-1/2} \cos \theta, \tag{6.8b}$$

or

$$A \simeq \frac{\cos^2 \theta}{2B} \, . \tag{6.8c}$$

The final result for x to leading order in B is

$$x = \frac{\sin \psi}{\cos \theta} + o(1) \, . \tag{6.9}$$

Further details of the expansion of $\tilde{x}$ and $\tilde{z}$ for $B \prec 1$ are left to the reader, see Exercise 6.1.

Next we proceed to the other limiting case of the interior problem, namely the *very wide tube*, $B \succ 1$. The leading term of the asymptotic expansion then corresponds to $B = \infty$. We may think of this as $l_g = \infty$ and thus get a different interpretation of the same problem. Since (6.5) does not contain x_d explicitly we may shift the origin to the left wall. As $l_g \uparrow \infty$ the previous origin recedes to infinity, and we have the case of a semi-infinite domain bounded at the left $(x_d = 0)$ by a wall. For further details, see Exercise 6.2.

The resulting problem is the same as the *exterior problem* whose solution is left to the reader (Exercise 6.3). We are led to the following rather curious result: *In two dimensions all exterior problems are the same (within trivial changes); the Bond number plays no role. The solution obtained corresponds to that for the limiting case of the interior problem as $B \uparrow \infty$.* This result is valid only for two dimensions (see Exercise 6.3); the first part of the preceding statement is closely related, in formulation and explanation, to the case $k = 0$ discussed in Section 5.

Case of Cylindrical Symmetry. We now let the inserted vertical tube be a circular cylinder. Nondimensional coordinates will be used; initially we shall use l_g as unit length, the coordinates are then defined by $(6.4b, c)$. Consistent with this we shall normalize the principal curvatures with l_g^{-1}, (but still use the old notation κ_j). Thus

$$\kappa_1 = -\cos \psi \, \frac{d\psi}{dr} = -\frac{d(\sin \psi)}{dr} \, , \quad \kappa_2 = -\frac{\sin \psi}{r} \, . \tag{6.10a,b}$$

Their sum is ("prime" now means $\frac{d}{dr}$)

$$\kappa_1 + \kappa_2 = -\frac{1}{r}(r\sin\psi)' = -\frac{1}{r}\frac{d}{dr}\left\{\frac{r\,z'}{[1+(z')^2]^{1/2}}\right\}. \qquad (6.10c)$$

Previously we had placed the zero level of the meniscus height (z_d or z) at infinite distance from the tube. For the *interior problem*[†] we shall introduce a parameter λ and a coordinate change

$$B\,z \to B\,z + \lambda, \qquad (6.11)$$

which allows us to choose a new zero level. Laplace's formula now reads

$$\frac{1}{r}\frac{d}{dr}\left\{\frac{r\,z'}{[1+(z')^2]^{1/2}}\right\} \equiv \cos\psi\,\frac{d\psi}{dr} + \frac{\sin\psi}{r} = B\,z + \lambda \qquad (6.12)$$

We introduce a new symbol D,

$$D = -\frac{\sin\psi}{r} + B\,z + \lambda = \cos\psi\,\frac{d\psi}{dr} = \frac{d\psi}{ds} = \sin\psi\,\frac{d\psi}{dz} = \text{ curvature of } z(r). \quad (6.13)$$

and rewrite (6.13) as a system of two equations with ψ as independent variable,

$$\frac{dz}{d\psi} = \frac{\sin\psi}{D}, \quad \frac{dr}{d\psi} = \frac{\cos\psi}{D}. \qquad (6.14a,b)$$

We now place the zero level of the meniscus height at the meniscus center. Boundary conditions for r and z are then

$$r = 0, \quad z = 0 \text{ at } \psi = 0. \qquad (6.14c,d)$$

The condition of the contact angle at the wall of the tube remains

$$r = 1 \text{ at } \psi = \psi_w, \quad i.e., \text{ at } \tan\psi = \cot\theta. \qquad (6.14e)$$

[†] For this problem we shall use various ideas, and some notation, from Concus (1968); further references are given there as well as discussions of some aspects of the meniscus problem not considered here, *e.g.* numerical results for values of B outside the range of applicability of perturbation methods.

The value of λ is, in principle, given as a function of B since there are three boundary conditions for two first-order equations which is possible only if the parameter λ is adjusted properly (see calculations below).

The difference in height between the center of the meniscus and the meniscus infinitely far away will be denoted by h (we still use l_g as unit length). We leave it to the reader, see Exercise 6.4, to show that with the choice of λ indicated above,

$$\lambda = B h . \tag{6.15}$$

Interior Problem. Narrow Tube. We shall now indicate how the actual calculations can be done for the interior problem; first we consider the *narrow tube* $(B \prec 1)$. It turns out that a usable asymptotic formula may be obtained with a regular perturbation technique. We assume r, z and λ to be regular power series in B

$$r(\psi) = r_0(\psi) + B r_1(\psi) \ldots , \tag{6.16a}$$

$$z(\psi) = z_0(\psi) + B z_1(\psi) \ldots , \tag{6.16b}$$

$$\lambda = \lambda_0 + B \lambda_1 \ldots . \tag{6.16c}$$

Putting $B = 0$ in (6.14) we find,

$$\frac{dr_0}{d\psi} = \frac{\cos\psi}{\lambda_0 - \frac{\sin\psi}{r_0}} , \quad \frac{dz_0}{d\psi} = \frac{\sin\psi}{\lambda_0 - \frac{\sin\psi}{r_0}} . \tag{6.17.a, b}$$

Solving, we find

$$r_0 = (2\sin\psi)\lambda_0^{-1} = \sec\theta \sin\psi , \tag{6.18a}$$

$$z_0 = 2(1 - \cos\psi)\lambda_0^{-1} = \sec\theta(1 - \cos\psi) , \tag{6.18b}$$

$$\lambda_0 = 2\cos\theta . \tag{6.18c}$$

The first expression for r_0 is obtained from the boundary condition at $\psi = 0$. From (6.18a) and (6.14e) we derive $\lambda_0 = 2\sin\psi_w = 2\sin\left(\frac{\pi}{2} - \theta\right) = 2\cos\theta$ which is (6.13c). From the value of z_0 at $\psi = 0$ we find (6.18b), with a general value of λ_0 and then with its value found above. The result may be interpreted from a geometrical and physical point of view, see Exercise 6.5.

Higher-order terms in (6.16) are found in a straightforward manner, see Exercise 6.6.

Interior Problem. Wide Tube. The interior problem for a narrow tube $(B \prec 1)$ has an asymptotic solution which is obtained naturally by a power series expansion in the small parameter B, and thus by a regular perturbation technique. On the other hand, for a *wide tube* $(B \succ 1, \epsilon \prec 1)$, qualitative reasoning suggests the use of singular perturbations. We may choose to think of this case as the case for which a weak surface tension is dominated by a strong gravitational force. Since curvature is maintained by capillary forces we expect the meniscus to be very flat, that is, $\psi = o(1)$ or equivalently $dz_d/dr_d = o(1)$; at least we expect these estimates to be valid over a large *core region*. However, since $\psi_w = O(1)$ (except when θ is very close to $\pi/2$) the core region cannot include the wall. As ψ decreases from the value ψ_w, gravity starts to dominate, but there must be a *boundary layer near the wall in which* $\psi = O(1)$. The boundary layer must merge with the core region, that is, the corresponding asymptotic solutions must match. Such matching expansions will justify and also make more precise the qualitative reasoning above.

We discuss first the *core region*. When $\psi = o(1)$ the term dz/dr $(= dz_d/dr_d)$ must be $o(1)$. Consider now Laplace's equation (6.12). To leading order the term $(z')^2$ drops out in the denominator of the curvature term. To obtain an equation as rich as possible we assume

$$r = O(\epsilon), \qquad z = O(\epsilon^2). \tag{6.19a,b}$$

The term λ is as yet undetermined; we assume $\lambda \precsim 1$ and discuss this later. Thus we use r^* as independent variable. Since we shall only compute the leading order term we do not introduce any notation for $\epsilon^{-1}z$ but put

$$z = \epsilon^2[f_0(r^*) + o(\epsilon^2)]\,. \tag{6.20}$$

The equation for the leading term f_0 is

$$\frac{1}{r^*}\frac{d}{dr^*}\left(r^*\frac{df_0}{dr^*}\right) = f_0 + \lambda\,, \tag{6.21a}$$

with boundary conditions

$$f_0(0) = 0\,, \qquad f'(0) = 0\,. \tag{6.21b, c}$$

The first condition places the zero level of the height of the meniscus at its center; the second condition (so far not used explicitly) states that the meniscus has a smooth stationary point at the center (a minimum in the present case). The smoothness implies that the solution to the homogeneous equation is a constant times the modified Bessel function $I_0(r^*)$. A particular solution of the inhomogeneous equation is simply the constant $-\lambda$. Thus the solution of (6.21) is

$$f_0 = \lambda[I_0(r^*) - 1]\,. \tag{6.22}$$

Taking the r^*-derivative and using (6.20) gives

$$\psi = \epsilon\lambda I_1(r^*)[1 + o(1)]\,. \tag{6.23}$$

The parameter λ will be determined by matching with the boundary layer solution. Note that above we have nowhere assumed that $\lambda = O_s(1)$.

At the wall of the tube $\psi = O(1)$; the same then holds in an adjacent boundary whose thickness is determined as follows: The order of ψ is the order of dz/dr; thus $1 - r$ and z are of the same order. The first principal curvature, that is, the curvature of the graph of z versus r was denoted by D in (6.13). We assume that now Bz dominates in D relative to the other terms which may be neglected; this may be justified from the assumption that dz/dr is now $O(1)$ rather than $o(1)$ as in the core. Bz is thus larger than before whereas λ remains the same; also it is intuitively plausible that, when B is large and $1-r$ is small, the other principal curvature is small. If Bz dominates, (6.14a) implies that $Bz^2 = O(\psi) = O(1)$. The order estimates just given imply that z and $1-r$ are $O(\epsilon)$. We do not introduce a special notation for variables of order unity but retain z and r. The calculations below will be carried out *to leading order only and the omission of higher-order terms will not be indicated explicitly.*

Using ψ as independent variable we get

$$\epsilon^2 \frac{dz}{d\psi} = \frac{\sin\psi}{z}, \quad z = 0 \text{ at } \psi = 0, \tag{6.24a, b}$$

$$\epsilon^2 \frac{d(1-r)}{d\psi} = -\frac{\cos\psi}{z}, \quad r = 1 \text{ at } \psi = \psi_w. \tag{6.25a, b}$$

These equations are derived from (6.14) using the simplifications and order estimates indicated above. The boundary condition for z is really a matching condition as will be discussed later. Solving (6.24) and (6.25) gives

$$z = 2\epsilon \sin\frac{\psi}{2}, \tag{6.26}$$

$$1 - r = \epsilon\left[\log\tan\frac{\psi}{4} + \cos\frac{\psi}{2} - \log\tan\frac{\psi_w}{4} - \cos\frac{\psi_w}{2}\right]. \tag{6.27}$$

The value of λ will now be determined by matching: The *core solution* $f_0(r^*)$, although valid for what is geometrically an inner region, is in our terminology the leading term of the *outer* expansion. For matching we must evaluate it for large values of r/ϵ. The asymptotic formula for I_0 gives

$$z = \epsilon^2\lambda\left(\frac{2\pi r}{\epsilon}\right)^{-1/2}\exp\left(\frac{r}{\epsilon}\right). \tag{6.28}$$

In the *boundary-layer solution*, that is, in the leading term of the inner expansion, consider large values of $\epsilon^{-1}(1-r)$, which means small values of ψ. Then $\tan\psi \sim \psi$

and $\cos \psi \sim 1$, and (6.27) give us ψ as a function of r and (6.26) then gives us z as a function of r in the matching region. Comparing this with (6.28) gives

$$\epsilon^{5/2} \lambda (2\pi)^{-1/2} r^{-1/2} \exp \frac{r}{\epsilon} = 4\epsilon \exp \left(\frac{-1}{\epsilon} \right) \tan \frac{\psi_w}{4} \exp \left[2 \cos \frac{\psi_w}{2} - 2 \right] \exp \frac{r}{\epsilon}. \quad (6.29)$$

The factor $\exp r/\epsilon$ cancels. Since $(1 - r) = o(1)$ in the matching region we may replace $r^{-1/2}$ by unity and the final result is

$$\lambda = 4\sqrt{2\pi} \, \epsilon^{-3/2} \exp \left(\frac{-1}{\epsilon} \right) \tan \frac{\psi_w}{4} \exp \left[2 \cos \frac{\psi_w}{2} - 2 \right]. \quad (6.30)$$

The method of matching as well as the result call for some comments. If we wanted to use Prandtl's simple matching recipe discussed in Section 1 we would try to equate the value of the core solution at $\epsilon^{-1}(1 - r) = 0$, that is, at the wall of the tube, with the value of the boundary-layer solution at $\epsilon^{-1}(1 - r) = \infty$. This method does not work here. However as pointed out in Section 1 it is a special case of intermediate matching which is the method used here. Note that we have not found it necessary to carry out the intermediate matching by using an intermediate variable. Still, we would like to verify, if not by proof but at least by consistency, that a required overlap domain exists (we do not need the maximal domain, just a plausible reason why the overlap domain is not empty). This and some other problems are left to the reader (Exercise 6.7), including the question of why we do not simply put the exponentially small λ equal to zero.

Exterior Problem. Narrow Tube. The exterior problem for very small Bond numbers is of special interest. The case shows similarities with problems discussed in Section 5; however, the switchback terms[†] are more complicated in the present

---

[†] Our calculations are based on those in Lo (1983), although our presentation of the results is different because we emphasize here different things. In her paper L. Lo alludes briefly to the interesting history of finding the switchback terms: Previous authors had used the method of matched asymptotic expansions but obtained results which they themselves recognized as faulty. Van Dyke insisted that the method should give correct results and guided L. Lo's research. She carried through some rather intricate calculations and obtained what now are considered to be correct answers.

problem. In order to exhibit the switchback terms we shall discuss higher-order approximations below.

It will be convenient to introduce a notation for $B^{1/2}$. We define

$$\eta = B^{1/2} = \frac{l_g}{l_c} . \tag{6.30a}$$

Thus

$$r^* = \frac{r_d}{l_c} = \frac{l_g}{l_c} \cdot \frac{r_d}{l_g} = \eta r . \tag{6.30b}$$

We may think of η tending to zero as the case of l_c fixed with l_g tending to zero. Thus in the limit $\eta = 0$ the tube has shrunk to an infinitely thin needle which has no influence on the fluid[†] so that the meniscus is flat. If a very narrow tube replaces the thin needle its influence if felt mainly in a small neighborhood with radius of order l_g (The *inner* region). The above considerations suggest the following. Away from the strong influence of the narrow tube (in the *outer region*) we use the capillary length as unit measure of the radial distance, that is, *we use r^* as outer variable.* We normalize the level of z_d, the height of the meniscus, to be zero at $r_d = \infty$ so that we dispense with the use of the parameter λ. As length scale for z_d we use l_g and use $z = z_d/l_g$ as nondimensional variable. The meniscus is no longer flat in the outer region, but at least $\psi = o(1)$ since $dz_d/dr_d = o(1)$ and $z = 0$ at $r^* = \infty$. At the wall of the tube, and hence near the tube $dz_d/dr_d = O(1)$. This is consistent with using r as inner variable and z as both outer and inner variable.[‡]

We shall now verify by considering the equations that the choice of variable

---

[†] This corresponds to the statement in Section 5 that a point has no cooling powers, see the discussion of Equation (5.5).

[‡] The exterior problem for small B shows certain similarities with the interior problem for large B. In both cases there is a large region ("outer" in the terminology of asymptotics) where $\psi = o(1)$, and a narrow inner region near the wall of the tube where $\psi = O(1)$.

made is appropriate. In *outer variables* the basic equation is

$$\frac{1}{r^*}\frac{d}{dr^*}\left\{\frac{r\frac{dz}{dr^*}}{\left[1+\eta^2\left(\frac{dz}{dr^*}\right)^2\right]^{1/2}}\right\} = z.$$ (6.31a)

The same equation expressed in *inner variables* is

$$\frac{1}{r}\frac{d}{dr}\left\{\frac{rz'}{\left[1+(z')^2\right]^{1/2}}\right\} = \eta^2 z, \qquad \text{``prime''} = \frac{d}{dr}.$$ (6.31b)

Assuming an *outer expansion*

$$z = f_1(r^*) + \eta^2 f_2(r^*) + o(\eta^2),$$ (6.32a)

we find

$$\frac{1}{r^*}\frac{d}{dr^*}\left\{r^*\frac{df_1}{dr^*}\right\} - f_1 = 0,$$ (6.32b)

$$\frac{1}{r^*}\frac{d}{dr^*}\left\{r^*\frac{df_2}{dr^*}\right\} - f_2 = \left(\frac{3}{2}f_1 - \frac{1}{r^*}\frac{df_1}{dr^*}\right)\left(\frac{df_1}{dr^*}\right)^2.$$ (6.32c)

An *inner expansion*

$$z = g_1(r) + \eta^2 g_2(r) + o(\eta^2),$$ (6.33a)

yields

$$\frac{rg_1'}{\left[1+(g_1')^2\right]} = \text{constant} = C, \qquad C = \cos\theta.$$ (6.33b)

$$\frac{1}{r}\frac{d}{dr}\left\{\frac{r}{\left[1+(g_1')^2\right]^{1/2}}\left[g_2' - \frac{g_1'g_2'}{\left[1+(g_1')^2\right]}\right]\right\} = g_1.$$ (6.33c)

The outer boundary condition for f_1 is $f_1(\infty) = 0$. Hence[†]

$$f_1 = C_1 K_0(r^*), \qquad C_1 = \text{constant} = \cos\theta.$$ (6.34)

[†] The same equation occurred in the core for the interior of the wide tube. There we wanted a smooth solution in the center. In the present case we do not care about a singularity at the center but want z to be zero at infinity. Hence we now use K_0 rather than I_0.

In (6.33b) we found the value of C from the inner boundary condition on the contact angle. In (6.24) we have given C_1 the same value; this is anticipating a result of matching.

The second term in the outer solution obeys the inhomogeneous equation (6.32c) and the condition $f_2(\infty) = 0$. L. Lo found a particular solution in the form of a rather complicated integral, but also observed that for matching we only need to evaluate it for small values of r^*. To the particular solution may be added a multiple of the homogeneous solution $K_0(r^*)$ which function also may be expanded for small r^*. Thus she finds the result

$$f_2 = \frac{c_2^3}{4}\left(\frac{1}{r^{*2}} - \log^2 r^* + \log r^* + \frac{3}{4}\right) + c_2\left(-\log\frac{\gamma r^*}{2}\right) + O(r^*). \qquad (6.35)$$

The constants will have to be determined by matching. In the course of matching we have to replace r^* by ηr; thus logarithmic switchback terms will arise as in Section 5. For this reason Lo uses several terms in the inner expansion, which is necessary if you insist on using a strict Poincaré expansion. However, as discussed in Sections 3 and 5, an alternative equivalent method is to use the form given in (6.33a) but let the constants of integration occurring in g_2 depend on η. We leave it to the reader to carry out the details for this alternative method (Exercise 6.8). The basic results are given in Lo (1965) and we warn the reader that, whichever method is used, the constant of integration will be unwieldy.

Exterior Problem. Wide Tube. This case is interesting, not only to complete the list of various asymptotic cases, but also because of its relation to various other problems; see Exercise 6.9.

Exercises for §6.

6.1. Consider the *interior problem in two dimensions*, $B \prec 1$. Find the two leading terms of the expansions in B of x_d and z_d. Find appropriate nondimensional variables, and discuss whether a regular or singular perturbation technique is most suitable. In particular, discuss the values of z_d at the center ($x_d = 0$) and at the walls of the tube ($|x_d| = l_g$); use $z_d = 0$ at $|x_d| = \infty$ as the zero level for the height of the meniscus.

6.2. *Interior problem, two dimensions but* $B \succ 1$. Discuss beginning of asymptotic expansion, in particular the value of the constant of integration A in (6.7a). Also discuss appropriate choice of nondimensional variables and the usefulness of a singular perturbation technique.

6.3. *Exterior problem, two dimensions.* Show that, while the interior of the tube is in pressure contact with the exterior, the width of the tube is irrelevant: Thus the Bond number is irrelevant; there is essentially *one* exterior problem, within a trivial coordinate change the descriptions of the meniscus to the left and to the right of the tube are irrelevant. Relate the exterior problem to the interior problem for the case $B = \infty$.

6.4. Prove equation (6.15).

6.5. For the interior problem with circular symmetry interpret the case $B = 0$ in terms of the gravitational force. From physics and geometry (Laplace's formula) find the curvature and derive (6.18).

6.6. Discuss the general forms of the terms in (6.16). Compute the coefficients of B.

6.7. For the matching leading to (6.30)

a) Exhibit some (not the maximal) overlap domain.

b) Justify (6.24b).

c) The formula shows that λ is exponentially small in ϵ. Why can we not simply put $\lambda = 0$?

6.8. Complete the determination of f_1, f_2, g_1, and g_2 of (6.32a) and (6.33a). Use the form suggested in the text.

6.9. Consider the exterior problem for B large. Find the leading term and discuss the second term. Compare with related problems.

2.7. Motion Past Two Newtonian Gravitational Centers

Introduction. A third example of an inner layer around a singular point will now be given; it is significantly different from the examples in the preceding two sections. We shall be concerned with motions of bodies in gravitational fields. The Newtonian laws of gravity will be assumed, furthermore the bodies will be represented by point-masses. They will create three-dimensional gravitational fields, although only planar motions will be considered here. As is well known, the *two-body problem*, that is the motion of two masses under the influence of their mutual gravitational attraction, is completely solved: It is easily reduced to the problem of a massive body moving around a fixed gravitational center; a first integral (constant of motion) is the total energy; a second integral gives the constancy of the angular momentum which implies that the motion is planar; furthermore a third integral, called the Runge-Lenz vector (the historical accuracy of that name may be argued) assures us that the motion is periodic. The heroic failures in trying to solve the corresponding *three-body problem* by finding integrals are described in the literature. Towards the end of the last century the suspicion arose that not enough integrals existed, even in principle. Poincaré introduced a simple version of the problem, called the *restricted three-body problem*: Two heavy bodies move in circular orbits, while the third one is a massless test particle, that is its mass is negligible compared to the two heavy masses and thus it moves in an a priori given gravitational field. By introducing a rotating system of coordinates the position of heavy bodies can be made fixed, at the cost of introducing Coriolis and centrifugal forces. Even the restricted problem turned out to be nonintegrable; however Poincaré's research on this problem is of fundamental importance for present-day dynamics.

With the advent of space flight the restricted three-body problem turned out to be of practical importance: The original and simplest example is the motion of a spaceship launched from the earth towards the vicinity of the moon, the influence of the sun and of Jupiter being cancelled to a fair accuracy by the orbiting of the earth-moon system determined by its surrounding gravitational field. The spaceship may be regarded as a test particle since its mass is negligible compared to that of the moon and thus, a fortiori, of the earth. Engineers involved in the earth-moon project spent considerable time on computing orbits, either by direct numerical computations or by using a patched-conic technique: The latter method assumed that initially the orbit was a Keplerian conic (ellipse) entirely governed by the gravitational field of the earth; then at a given moment the gravitational field of the moon suddenly took over, after which the spaceship's orbit was entirely governed by the field of the moon. It was suggested by Lagerstrom and Kevorkian (1963b) that the problem could be treated more systematically and efficiently by a singular-perturbation technique using the fact that the ratio of the mass of the moon to that of the earth is small, approximately equal to .01. To try out the technique and to explain it they first studied a simpler problem, namely a test particle moving under the influence of two fixed gravitational centers. The results were published in Lagerstrom-Kevorkian (1963a) and will be discussed below. By "fixed centers" we mean "fixed in an inertial frame", that is there are no Coriolis or centrifugal forces. Euler found an integral for this problem, thereby solving it in principle.[†]

[†] This problem may seem to be unphysical (how can the masses remain in a fixed position?) and the solution mainly an example of Euler's analytic virtuosity. However, the Coulomb field is mathematically the same as the Newtonian field and the attraction of each center may be of arbitrary magnitude and sign. This makes it possible to derive various interesting limiting cases: The Runge-Lenz vector is a limiting case of the Euler integral; so is the integral of the Newtonian center in a uniform gravitational field, a problem in classical dynamics which in quantum mechanics leads to the Stark effect. The Euler solution of the problem of two fixed centers, including the integral of motion, is given in Whittaker (1904, 1940) but unfortunately dropped (together with several interesting lim-

Here we shall discuss the fixed-center problem; the basic ideas of the perturbation technique will be the same as in the restricted three-body problem but the analytical computations will be simpler.

Basic Equations of the Problem. We shall go directly to nondimensional variables; masses will be normalized with the mass m_e of the heavier particle (called "earth" for convenience), distances with the fixed distance D between the two centers, and time with a suitable dimensional time involving the gravitational constant G (see Exercise 7.1). We use Cartesian coordinates x, y for the position of the test particle, and place the heavier particle ("earth") at the origin and the lighter particle ("moon") of mass ϵ ($\epsilon \prec 1$) at the point $x = 1$, $y = 0$. The equations of motion are then,

$$\ddot{x} = \frac{d}{dx}\left(\frac{1}{r} + \frac{\epsilon}{r_m}\right), \quad \ddot{y} = \frac{d}{dy}\left(\frac{1}{r} + \frac{\epsilon}{r_m}\right), \qquad (7.1a,b)$$

where $r^2 = x^2 + y^2$, $r_m^2 = (x-1)^2 + y^2$, and "dot" denotes the derivative with respect to time.

There are two integrals of motion. The obvious one is the total energy (Hamiltonian),

$$\text{Energy: } \frac{1}{2}\left(\dot{x}^2 + \dot{y}^2\right) - \frac{1}{r} - \frac{\epsilon}{r_m} = H. \qquad (7.2)$$

The second one is the integral found by Euler, which we give here in a somewhat more transparent form than that given by Whittaker,

$$\text{Euler Integral: } LL_m + \frac{x}{r} - \frac{\epsilon(x-1)}{r_m} = J, \qquad (7.3a)$$

where L and L_m are the angular momenta relative to the position of the earth and the moon respectively,

$$L = x\dot{y} - y\dot{x}, \quad L_m = (x-1)\dot{y} - y\dot{x} \equiv L - \dot{y}. \qquad (7.3b,c)$$

---

iting cases) in many recent textbooks on analytical dynamics; however, a brief treatment is given in Corben-Stehle (1960). Fokas (1979) and Fokas and Lagerstrom (1980) made a systematic study of Hamiltonian systems with integrals quadratic in the momenta, in particular those due to several fixed centers and their limiting cases.

The constancy of J is easily verified by computing its time derivative. Thus we have a planar problem with two degrees of freedom and two independent integrals; if we consider three-dimensional motion we use the angular momentum with respect to the x-axis as a third integral. In either case the problem is thus in principle integrable by Liouville's theorem of integrability. However, in the restricted three-body problem no analog of the second integral exists. Therefore we shall not make use of the second integral in the perturbation calculations below. The discussion will be based on that of Lagerstrom-Kevorkian (1963a).

Qualitative Description of Orbits. Since $\epsilon \prec 1$ it is clear that when the orbit of the particle is not too near the moon the influence of the earth dominates; an orbit which never comes close to the moon can be described by classical perturbation methods developed in celestial mechanics. The terms "near" or "close" are of course vague; it will be seen that a critical value of the order of nearness is $O_s(\epsilon)$. Here we shall consider orbits which pass within a distance of order ϵ to the moon and which start near the earth in such a way that $y = O(\epsilon)$ during the approach to the neighborhood of the moon. This approach can then be described by an expansion whose leading term is Keplerian relative to the earth; only higher order corrections will show the influence of the moon. We shall, somewhat arbitrarily, assume that the motion is elliptic which means that $H < 0$; on the other hand for the orbit to reach the moon we must assume that $H \geq -1$. This description of the approach leg will be part of an *outer expansion* using x, y, and t as *outer variables*. The moon passage will be described by an *inner expansion* using *inner variables* obtained by scaling $x - 1$, y, $t - t_m$ by ϵ where t_m is a suitable time (in outer variables) of arrival near the moon. Matching will require that the outer x-velocity at $x - 1 = 0$ equals the inner velocity at $\tilde{x} = -\infty$; the inner expansion of the moon passage will then yield a hyperbola with possible higher-order corrections. (Note that an elliptic approach orbit was *chosen* as a convenient interesting case; the use of the perturbation methods *forces* us to assume a hyperbolic moon passage.) After the moon passage the outer expansion will describe another orbit elliptic relative to the earth. Viewed with the scale of the outer expansion the moon passage will appear as a point. At this point the magnitude of the velocity is continuous while its direction will change discontinuously, corresponding to the change from the initial asymptote to the final asymptote of the hyperbolic moon passage. The velocity at the moon will determine the second leg of the outer expansion; here the value of y need no longer be $O(\epsilon)$.

We shall now discuss some analytical details of an asymptotic description of the complete orbit.

Outer Expansion for Approach Leg. Below it will be understood that terms of order ϵ^2 are neglected even if the $O(\epsilon^2)$ is not written explicitly. Since $y = O(\epsilon)$ by assumption, it is convenient to use x rather than t as the independent variable and we assume outer expansions of the form

$$y = \epsilon y_1(x) , \qquad t = t_0(x) + \epsilon t_1(x) . \tag{7.4}$$

The energy integral is now

$$H_0 + \epsilon H_1 = \frac{\dot{x}^2}{2} - \frac{1}{x} - \frac{\epsilon}{x-1} = \frac{1}{2 \left(t_0' + \epsilon t_1' \right)^2} - \frac{1}{x} - \frac{\epsilon}{x-1} , \tag{7.5}$$

where "prime" denotes d/dx. To leading order

$$\frac{2^{-1/2}}{t_0'} = \left(\frac{1}{x} + H_0 \right)^{1/2} \equiv \left(\frac{1}{x} - A^2 \right)^{1/2} , \qquad A^2 = -H_0 . \tag{7.6}$$

(As stated earlier $-1 \le H_0 < 0$.) This equation may be integrated assuming $t(0) = 0$,

$$\sqrt{2} A^3 t_0 = \arcsin \left(A^2 x \right)^{1/2} - \left[A^2 x \left(1 - A^2 x \right) \right]^{1/2} . \tag{7.7}$$

In particular t_m, the time of arrival at the moon, is to leading order

$$t_m = t_0(1) = 2^{-1/2} A^{-3} \left[\arcsin A - A \left(1 - A^2 \right)^{1/2} \right] . \tag{7.8}$$

Without using the formula for t_0 we find directly from the energy integral that the absolute value of the x-component of the velocity at the moon is, to leading order

$$\left. \left| \frac{dx}{dt} \right| \right|_{x=1, t=t_m} = 2^{1/2} \left(1 - A^2 \right)^{1/2} . \tag{7.9}$$

Here we have again assumed that y, and hence $\dot{y}$, are of order ϵ.

To complete the outer expansion for the approach leg we consider now the y-coordinate. Since by assumption y is of order ϵ we put

$$y = \epsilon y_1(x) . \tag{7.10}$$

The equation for y_1 is easily found and has a relatively simple general solution. However, here we shall not use a solution based on initial conditions near the earth ($x = 0$). Instead we assume an orbit such that

$$y_1(1) = -C , \qquad C = \text{positive constant} . \tag{7.11}$$

(The assumption of a negative value has no special significance, it only serves to fix the ideas.) The meaning of (7.11) will be obvious below when we consider matching.

Inner Expansion for Moon Passage. As inner coordinates we use

$$\tilde{x} = \frac{x - 1}{\epsilon} , \qquad \tilde{y} = \frac{y}{\epsilon} , \qquad \tilde{t} = \frac{t - t_m}{\epsilon} , \qquad \tilde{r} = \frac{r_m}{\epsilon} . \tag{7.12}$$

The justification of the choice of scaling is left to the reader (Exercise 7.4). In inner variables the motion is to leading order a Keplerian orbit relative to the moon. The elements of the orbit are found by matching with the outer expansion for the approach. We use the simple recipe of Section 1: The outer velocity at $x^* = 0$ ($x = 1$) matches with the inner solution at $x^* = -\infty$ (Exercise 7.5). The y-component at $x = 1$ does not contribute to L_m. Hence we get from (7.9), (7.11) and matching that at $x = 1$,

$$L_m = \epsilon C \left[2 \left(1 - A^2 \right) \right]^{1/2} . \tag{7.13}$$

Since L_m is constant during the moon passage, (7.13) gives one element of the orbit.

The total energy H is a constant for the entire motion. We need to interpret this for the moon orbit: In inner coordinates the kinetic energy and the potential energy due to the moon are of order unity. Since $x = \epsilon \tilde{x} + 1$, $y = \epsilon \tilde{y}$ we find that $1/r = 1 + O(\epsilon)$. Thus to leading order the earth's potential simply contributes a constant. While H is a constant, H_m the total energy of the moon passage can thus conveniently be written

$$H_m = H + 1 = 1 - A^2 + O(\epsilon) > 0. \qquad (7.14)$$

Since $A < 1$ we have verified that the moon passage is hyperbolic. We now know the values L_m, H_m and the position of the approach asymptote. By standard formulas we find the position of the final asymptote, in particular the angle between the two asymptotes.

Outer Expansion of the Second Leg. After the moon passage the motion is again Keplerian relative to the earth. By the same principle of matching as before, the final asymptote of the moon passage gives us the initial values of the velocity components at $x = 1$, $y = 0$. The value of y need no longer be of order ϵ.

Combining the outer expansion of the approach leg and the final leg we find an orbit which is continuous at $x = 1$, $y = 0$ but whose derivative (direction of velocity) changes discontinuously there. This discontinuity is smoothed out by a derivative layer (compare Section 4) which in fact is the hyperbolic motion around the moon described in inner coordinates.

Comments. As in the preceding Sections 5 and 6, the inner layer occurs at a singular point (for Section 6 only for the exterior problem at small Bond numbers). However, in those problems we considered a finite size object (circle, sphere) which shrinks to a point in the outer limit. In the present problem the replacement of the heavy bodies by point masses has nothing to do with the particular perturbation problem studied here. The use of perturbation techniques is justified by the assumption that one mass is much smaller than the other; singular perturbation techniques are appropriate only when a portion of the orbit is such that the closeness to the moon gives a non-negligible force in spite of smallness of the mass of the moon

The discussion here has been sketchy. Further details, such as higher order approximations and composite expansions, may be found in Lagerstrom-Kevorkian (1963a) (note that there the mass of the earth is assumed to be $1 - \epsilon$ rather than 1 and that special initial conditions were used). The more complicated restricted three-body problem is discussed by Lagerstrom and Kevorkian (1963b) and (1963c); the second paper discusses the case of orbits with minimal energy. Kevorkian initiated the use of the singular-perturbation method for finding periodic orbits, see Kevorkian-Lancaster (1968) for the restricted three-body problem and Kevorkian-Cole (1980); both publications contain various pertinent further references which are not listed here.

Exercises for §7.

7.1. Discuss in detail the dimensional analysis mentioned just before equation (7.1).

7.2. Show that for the approach leg $\dot{L} = O(\epsilon^2)$.

7.3. Find the outer equation for y_1 and its general solution.

7.4. Justify the choice of inner coordinates.

7.5. Discuss the matching of the inner and outer expansions in detail and justify the matching recipe used in the text.

7.6. Find a suitable limit process by which the problem of two fixed centers turns into the problem of one center of mass $\eta \prec 1$ is located at the origin and the other center gives rise to a uniform gravitational field whose acceleration due to gravity is of unit strength in the negative x-direction. What happens to the Euler constant of motion (7.3)?

7.7. For the potential mentioned above, consider a particle of negligible mass which
at $t = 0$ has zero velocity and position $x = 1$, $y = B\eta$. What happens in
the neighborhood of $x = 0$, $y = 0$?

2.8. Relaxation Oscillations

An oscillatory function which alternates between slow changes and relatively rapid
changes is said to describe a *relaxation oscillation*; such a phenomenon is interesting
both from a theoretical and from a practical engineering point of view. We shall
restrict ourselves to one-dimensional periodic functions of one scalar variable (which
may be interpreted as "time" or as "space"); the method of matched asymptotic
expansions is then often suitable for obtaining an approximate quantitative description of the oscillation. Below we shall discuss a few examples; only one of those, the
van der Pol relaxation oscillation, has been treated extensively in the literature.

A Hamiltonian Equation. The equation is

$$\epsilon u'' + u' + u = 0 . \tag{8.1}$$

It differs from (4.1a) only by the sign of the third term; many ideas of Section 4[†]
may be used, in particular the derivation of the equation from a Hamiltonian, but
some essential differences of the solutions will be introduced by the change of sign.
The leading term of the inner solution will be taken unchanged from Section 4 (since
the sign of the third term does not matter); one essential change is that the previous
outer solution $f_0 = x +$ constant now has to be replaced by

$$f_0 = -x + \text{constant} . \tag{8.2}$$

[†] It is left to the reader to fill in various details of the discussion below; see Exercise 8.1.

Another difference, essential for the possibility of periodic solutions, is that the singular point has now changed from a saddle point to a center; this is seen by introducing the Kaplun scaling as in Section 4 and constructing the universal phase portrait in analogy with Figure 4.1.

Now let $\epsilon \prec 1$ and consider the closed curves far away from the center. Interpreted in the original variables, that is, as functions u of the variable x, the outer solutions corresponding to such functions are sawtooth shaped: For some positive number b the variable u decreases with slope minus one from $u = b$ at $x = -b$ to $u = -b$ at $x = b$ (we are free to choose the position of the origin of the x-axis) where it increases discontinuously to $u = b$ again; this behavior is repeated with period $2b$ over the entire x-axis. The discontinuous upjump is smoothed out by an inner solution

$$g_0 = b \tanh \frac{b\tilde{x}}{2} \,, \tag{8.3a}$$

where

$$\epsilon \tilde{x} = x - x_d \,, \qquad x_d = 2nb \,, \qquad n = \text{integer} \,. \tag{8.3b}$$

A solution uniformly valid to order unity in ϵ may be found as in Section 4.

Thus, an analytical leading-order asymptotic solution is easily found; here we shall not pursue the discussion of (8.1) further or discuss possible physical interpretations.

The van der Pol Equation. The equation may be written

$$\frac{d^2u}{d\tau^2} + \mu(u^2 - 1)\frac{du}{d\tau} + u - 0 \,, \qquad \mu > 0 \,. \tag{8.4}$$

In applications τ denotes a nondimensional time. It does not describe a conservative system, but an oscillator which gains energy for $|u| < 1$ and loses energy for $|u| > 1$. It is related to various other equations, is of interest in physics and engineering, and is often used to illustrate the phenomenon of limit cycles and relaxation oscillations. The equation has been studied extensively from various mathematical points of view; J. D. Cole, see e.g., Cole (1968), seems to be the first one to study the application of various systematic asymptotic methods to extreme values of μ, including the approach to a limit cycle for $\mu \prec 1$. We shall here briefly discuss limit cycles for $\mu \succ 1$, which is the case of relaxation oscillations. Since there is a vast literature on this subject and since the calculations seem to become by necessity rather involved, we shall refer to Kevorkian-Cole (1981) for details and give mostly a qualitative outline here.[†]

For a treatment of (8.4) by layer-type techniques it is convenient to transform (8.4) into

$$\epsilon\frac{d^2u}{dt^2} + (u^2 - 1)\frac{du}{dt} + u = 0\,,\qquad(8.5a)$$

where

$$t = \frac{\tau}{\mu}\,,\qquad \epsilon = \frac{1}{\mu^2}\,.\qquad(8.5b,c)$$

Assume now $\epsilon \prec 1$. To describe the limit cycle qualitatively let us consider a value of u which is > 1 but $u - 1$ is not very small. There is no reason for the second derivative to very large; hence we neglect the first term and use the outer equation obtained by putting $\epsilon = 0$ in (8.5). We see that du/dt is negative and that u decreases with time. If we still use the outer equation, we find that du/dt assumes very large negative values and reaches the value $-\infty$ at a finite time which we may give the value $t = 0$ (the equation is autonomous). The general solution for f_0, the leading term of the outer solution, is

$$\log|f_0| - \frac{f_0^2}{2} = t - t_0\,.\qquad(8.6)$$

Above we chose as constant of integration $t_0 = 0$. The outer solution can no longer continue along the branch of (8.6), since f_0 cannot be positive for $t > 0$. Since

---

[†] The reference just given describes a problem from electrical engineering, which leads to van der Pol's equation; its relation to Rayleigh's equation is also indicated there.

df_0/dt is $-\infty$ at $t = 0$, $f_0 = 1$, we expect a very rapid down jump to a branch of the outer solution with $f_0 < -1$. In the outer solution, the jump will appear as a discontinuity to be smoothed out by an inner solution. The inner variable has to be chosen so that the second derivative is restored. This is achieved by choosing

$$t^* = \frac{t - \delta(\epsilon)}{\epsilon} ,$$
(8.7)

where the displacement $\delta(\epsilon)$ has to be determined (see comments below). The equation for the leading term is then

$$\frac{d^2 g_0}{dt^{*2}} - \left(1 - g_0^2\right) \frac{dg_0}{dt^*} = 0 ,$$
(8.8)

with the first integral

$$\frac{dg_0}{dt^*} - g_0 + \frac{1}{3} g_0^2 = C_0 .$$
(8.9)

Matching with the outer solution at $t^* \to -\infty$, $t = 0$, $f_0 = 1$ gives

$$C_0 = -\frac{2}{3} .$$
(8.10)

As $t^* \to +\infty$, $g_0 = -2$. From this we conclude that the downjump in the outer solution goes from $f_0 = 1$ to $f_0 = -2$. Thus, at $t = 0$, the jump overshoots the region of negative damping to a point $f_0 = -2$ in the region of positive damping. Then the previous argument applies. The outer solution takes over and the solution approaches the value $f_0 = -1$ from below. At this point, the outer solution has a discontinuous upjump as before. The leading order of the outer and inner solutions has thus been found. The reasoning is taken from Cole's book where the next terms are found and $\delta(\epsilon)$ is determined. For a discussion of the related Rayleigh's equation, see Kevorkian-Cole (1981).

The Equation of Fitz Hugh and Nagumo. This equation is meant as an approximation to the Hodgkin-Huxley equation for the propagation of nerve impulses (voltage
pulses); see Hodgkin and Huxley (1952), Fitz Hugh (1961), Nagumo et alia (1975).
The perturbation analysis below will be based on that of Casten, H. Cohen and
Lagerstrom (1975) (this paper also gives further references to the literature and a
discussion of some background from chemistry and biology).

The approximating equation for the propagation of the voltage v is, in nondimensional variables,

$$\frac{\partial v}{\partial t} + f(v) + z = \frac{\partial^2 v}{\partial x^2}, \qquad \frac{dz}{dt} = \epsilon v, \tag{8.11a,b}$$

where $f(v) = v(v-a)(v-1)$, $0 < a < 1$, and ϵ is a small positive parameter.
We shall discuss solutions which depend only on $\xi = x + \theta t$, $\theta = $ constant; it can
be shown that without loss of generality we may assume θ to be positive so that
the solutions represent waves travelling to the left. In this case (8.11) reduces to

$$\frac{d^2 v}{d\xi^2} - \theta\frac{dv}{d\xi} - f(v) - z = 0, \qquad \theta\frac{dz}{d\xi} = \epsilon v. \tag{8.12a,b}$$

Perturbation methods will show that each value of θ, at least in a certain range,
corresponds to a value of the amplitude. In the limit $\epsilon \downarrow 0$

$$v(\xi, \epsilon) \to v_0(\xi), \qquad z \to K = \text{constant}, \tag{8.13}$$

so that in this limit (8.12) reduces to the autonomous second-order equation

$$\frac{d^2 v_0}{d\xi^2} - \theta\frac{dv_0}{d\xi} - f(v_0) - K = 0. \tag{8.14a}$$

As will be seen later ξ is actually an inner variable and thus the statements just
made are not uniformly valid. As an alternative notation we shall also express the
last two terms of (8.14a) by

$$f(v_0) + K = (v - V)(v - W)(v - U), \tag{8.14b}$$

where $V < W < U$. Solutions for travelling waves will be possible only when V and U, and hence also W, are real. For $K = 0$, the values of the roots are $V = 0$, $W = a$, $U = 1$. The position of the roots as K varies are easily seen qualitatively from the graph of the cubic $f(v_0)$ vs. v_0 (see Casten, H. Cohen, Lagerstrom, p. 393).

Upjump for a Solitary Wave. Experimental evidence indicates that there are waves for which $v = 0$ at $\xi = -\infty$, then v rises slowly, suddenly rises fast to a positive value of order unity, varies slowly, descends fast to a value which is slightly negative and finally slowly approaches the value $v = 0$ as ξ tends to very large values. Such a wave is called a solitary wave; it may be regarded as a limiting case of periodic motion as the period tends to infinity. If (8.11) is to qualify as an approximating equation for voltage propagation it must have a solution which describes solitary waves approximately; actually the solution should satisfy (8.12).

To find such a solution we start with (8.14) and try first the value $K = 0$. We consider the Poincaré phase portrait in the plane with coordinates v_0 and $w = dv_0/d\xi$. There are three singular points, all of which have $w = 0$. The values of v_0 are $v_0 = W$ (center) and $v_0 = V$ or $v_0 = U$ (saddlepoints). There are two heteroclinic solution curves, v_{0u} going from $(0, V)$ to $(0, U)$ with v_0 positive and v_{0d} returning through negative values of v_0. To find analytic expressions for the heteroclinic orbits we replace $dv_0/d\xi$ by w, $d^2v_0/d\xi^2$ by $w\,dw/dv_0$ and look for solutions for which w is a polynomial of v_0. This is possible only if w has the form

$$w = \lambda(v_0 - V)(U - v_0), \qquad \lambda = \pm\frac{1}{\sqrt{2}}, \qquad (8.15a)$$

and

$$\theta = \lambda G(U), \qquad G(U) = U + V - 2W. \qquad (8.15b)$$

For a given θ there are two possibilities: A) $\lambda > 0$ and

$$v_{0u}(\xi) = \frac{U \exp(h\xi) + V}{1 + \exp(h\xi)}, \quad \sqrt{2}h = U - V, \tag{8.16a}$$

B) $\lambda < 0$ and

$$v_{0d}(\xi) = \frac{U \exp(-h\xi) + V}{1 + \exp(-h\xi)}. \tag{8.16b}$$

(The subscripts "u" and "d" denote "upjumps" and "downjumps," terms which will be explained later.)

We see that the assumption $K = 0$ is reasonable for the beginning of v_{0u} as given by (8.16a) with $V = 0$ and $U = 1$. The voltage v is exponentially small in an interval semi-infinite in ξ; thus it is reasonable to assume $z = K = 0$ for ξ negative and also for the "upjump" where it rises in an interval of order unity near the origin. However, after the upjump v_{0u} is near unity and increases to that value on $\xi \to \infty$; it is then no longer reasonable to assume that the integral of $dz/d\xi$ remains of order ϵ. Furthermore, the solution v_{0u} does not give the downjump which we had expected. Without looking at the details of the analytic solution we know from general theory (linearized theory near the singularity) that the solution reaches the saddle-point $(1,0)$ only at $\xi = \infty$; in particular, $w = 0$ at the saddle point and very small near it. This last formulation of the difficulty makes its resolution obvious. In the case of the van der Pol relaxation oscillation a solution for which the velocity tended to infinity in absolute value necessitated the introduction of a scaled up variable, that is a variable which was inner relative to the previous outer variable. Similarly, a vanishing value of w indicates that we should regard ξ as an inner variable relative to an outer variable which we shall call η. The assumption that η should be scaled by ϵ is made natural by (8.12b). Thus we define

$$\eta = \epsilon\xi, \quad u(\eta;\epsilon) = v(\xi;\epsilon), \quad u_0(\eta) = u(\xi;0). \tag{8.17a,b,c}$$

Outer Solution. Rewriting (8.12) in terms of the variables u and η gives

$$\epsilon^2 \frac{d^2 u}{d\eta^2} - \epsilon\theta \frac{du}{d\eta} - f(u) - z = 0 , \tag{8.18a}$$

$$\theta \frac{dz}{d\eta} = u . \tag{8.18b}$$

The equation for the leading term in the outer expansion is then

$$-\theta f'(u_0) \frac{du_0}{d\eta} = u_0 , \quad \theta \frac{dz_0}{d\eta} = u_0 , \tag{8.19a, b}$$

with the general solution (the additive constant of integration is hidden in η)

$$\frac{\eta}{\theta} = -\frac{3}{2}u_0^2 + 2(1 + a) - a \log|u_0| . \tag{8.20}$$

(The value of z_0, the leading term of z, will be discussed below.) Hence u_0 becomes a multi-valued function of η. To see this we observe that the cubic function $f(u_0)$ has two stationary points, a maximum at $u_0 = u_{\max}$ and a minimum at $u_0 = u_{\min}$ where

$$0 < u_{\max} < u_{\min} < 1 . \tag{8.21}$$

At those values $du_0/d\eta$ is infinite; the two infinities in the slope and the logarithmic infinity divide the graph of u_0 into four branches. The branches need not be connected; a discontinuous jump from one branch to another may be possible, a necessary condition being that it can be smoothed out by an inner solution. In each branch used η can be shifted by an additive constant of integration of (8.19). We can now interpret the infinitely slow approach, $v_{0u}(\xi) \to 1$ as $\xi \to \infty$, as a matching condition. The outer solution $u_0(\eta)$ attains the value unity only in the uppermost branch for which $u_0 > u_{\min}$. We may shift the value of η such that $u_0 = 1$ at $\eta = 0$. The matching condition fulfilled is then

$$\lim_{\eta \downarrow 0} u_0(\eta) = v_0(\infty) = 1 . \tag{8.22}$$

We may assume that $z = K = 0$ at $\eta = 0$. Then as u_0 decreases with increasing η, z increases according to (8.19b),

$$z_0 = \frac{1}{\theta} \int_0^\eta u_0(\eta)\, d\eta \, . \qquad (8.23a)$$

However, it is more convenient to treat z_0 as a function of u_0 and we find from (8.18a) for $\epsilon = 0$,

$$z_0 = -f(u_0) \, . \qquad (8.23b)$$

Downjump. The use of an outer solution resolved the contradiction between the original assumption $z = \text{constant} = 0$ and the fact that it seemed necessary for z to increase because of the slow variation of v with ξ near the singular point. So far we have obtained a mathematical description of the first part of a solitary wave: the slow rise from $v = 0$, described by an exponentially small function of the inner variable ξ; the fast rise to $v = 1$ in an interval of order unity on ξ; the slow decrease of v described by a function of the outer variable η. According to our qualitative description of the solitary wave the outer solution should be followed by a rapid decrease ("downjump"), a phenomenon which we have not found mathematically. However, the outer solution actually requires a downjump: we assume that the origin of the η-axis has been fixed as in the matching condition (8.22); as η increases from zero $f'(u_0)$ is positive and hence u_0 decreases, eventually reaching the value $u_{\min}$ where the derivative $du_0/d\eta$ equals $-\infty$. As was the case for the van der Pol equation the outer solution must jump discontinuously to a lower value. From the outer equation we can determine that this lower value must occur on the lowest branch of the solution for u_0, that is u_0 must be negative. As η increases towards $+\infty$, the value of u_0 increases to zero. So far we have shown how one can obtain some bounds for the beginning and the end of the downjump; to find more precise values we need to study (8.14). For the upjump we had assumed that $K = 0$; however, after the upjump K increases to a value so far unknown. We expect a heteroclinic orbit which in the phase-plane goes from $(U,0)$ to $(V,0)$. From (8.14) we see that the pair V and U determines K and K determines V and U. The essential idea to save us from a vicious circle is that the value of θ must be the same for the downjump and the upjump. Thus from (8.15) and (8.16) we can determine the downjump, that is the constants of (8.16b) and the constant of integration inherent in ξ. The downjump takes u_0 from the top branch where $u_0 > u_{\min}$ to the bottom branch where $u_0 < 0$. In the bottom branch u_0 increases with η, reaching the value zero at $\eta = \infty$. For details see Casten, H. Cohen and Lagerstrom (1975). Some details are left to the reader in Exercise 8.2.

The solitary wave seems to be the most interesting case for biologists. However, for the simplified equation studied here the mathematical methods used also give relaxation oscillations with finite period. A key idea in obtaining solutions is again the obvious requirement that the wave velocity θ at the upjump must equal that at the downjump.

Exercises for §8

8.1. a) Consider all variations of Equations 8.1 obtained by keeping the first term unchanged and changing the first and/or the second plus sign into a minus sign. Show that within simple coordinate changes involving real constants there are essentially two inequivalent equations, (8.1) and (4.1a). Show that these two equations are equivalent if complex variables are allowed.

b) Using the Kaplun scaling of Section 4, find an integral of (4.1), express that equation as a Hamiltonian system, and describe the universal phase portrait.

c) Using ideas from Section 4, fill in the details in the discussion of (8.1) in the present section.

8.2. a) Sketch u_0 vs. η for the solitary wave from (8.20). Assume $0 < a < 1/2$. Pay special attention to points with vertical tangents. Draw the four disconnected curves, each single valued. Sketch the downjump as a discontinuity in η.

b) For the solitary wave calculate K for the downjump and sketch the phase portrait of v_{0d}.

8.3. Describe in words the periodic phase portrait in (v_0, w, z)-space which corresponds to a travelling wave with finite period.

2.9. Notes on Modulation Techniques: Stokes-Lindstedt, Multiple Scales, Averaging

This book deals almost exclusively with layer-type techniques. With some oversimplification one may state that the basic techniques and ideas were developed during the 1940's and 1950's with Prandtl's boundary-layer theory as a starting point. In the nineteenth century astronomers encountered another phenomenon of nonuniformity, namely secular terms. Perturbation solutions were found which appeared valid for shorter intervals, say a couple of years; however, for time intervals of the order of magnitude of a century the secular[†] terms occurring were unrealistically large in absolute value. It was realized that the secular terms expressed a resonance which was not a physical phenomenon but a fake resonance due to bad mathematics. Methods were developed to obtain approximations which avoided secular terms and were valid for long time intervals. These methods were later applied further in the study of engineering problems of oscillations, electrical or mechanical. We shall here give some simple classical examples of the mathematics involved, and then give some reasons why we do not pursue this subject further in this book.

Periodic Motion; Lindstedt's Method. Consider an undamped oscillatory mass-spring system with a slightly nonlinear hard spring. With a suitable normalization

---

[†] From the latin word for century.

the governing equation may be written

$$\frac{d^2_u}{dt^2} + u + \epsilon u^3 = 0, \qquad 0 \leq \epsilon \prec 1.$$ (9.1)

A straightforward expansion procedure would assume an expansion of the form

$$u \simeq u_0(t) + \epsilon u_1(t) + ..$$ (9.2)

If we choose $u_0 = A \cos t$ the equation for u_1 is

$$\frac{d^2 u_1}{dt^2} + u_1 = -\frac{A^3}{4}(3 \cos t + t \cos 3t).$$ (9.3)

The response to $\cos 3t$ is an oscillation of three times the basic frequency. However, the first forcing term gives resonant response $-\frac{3}{8}A^3 t \sin t$. This is a secular form: the amplitude of each oscillation increases indefinitely with time. Also, the motion should be periodic, but $u_0 + \epsilon u_1$ is not a periodic function. In this simple example we see immediately what is wrong with (9.2). The motion is period and can thus be developed in a Fourier series in the basic period. The cubic term changes the period from its value of $\epsilon = 0$. If we develop $\cos(1+\epsilon)t$ in ϵ we get $\cos t - \epsilon t \sin t+$, which shows how a wrong period gives a secular term. Thus the reason for the secular terms is clear; we now consider methods for avoiding them.

One method is that of Lindstedt† since the frequency depends on ϵ and its value for $\epsilon = 0$ is unity we assume

$$\omega = 1 + \epsilon \omega_1 + \epsilon^2 \omega_2 \ldots ,$$ (9.4a)

and use a modified time scale defined by

$$t^* = \omega t.$$ (9.4b)

---

† This name seems acceptable although "Stokes-Lindstedt" would also be acceptable; Stokes (1847) had used a similar method earlier, but Lindstedt developed this method probably independently of Stokes. The second volume (1893) of Poincaré's "Les Méthodes Nouvelles de la Mécanique Céleste" has as subtitle "Methods of Newcomb, Gyldén, Lindstedt and Bohlin. In particular, Poincaré made various contribution to the study of Lindstedt's method.

We introduce t^* into and replace (9.2) by

$$u \sim u_0(t^*) + \epsilon u_1(t^*) + \dots \quad , \tag{9.5}$$

and find

$$u_0 = A \cos t^* , \tag{9.6a}$$

and

$$\frac{d^2 u_1}{dt^{*2}} + u_1 = \left(-\frac{3A^3}{4} + 2\omega_1 A \right) \cos t^* - \frac{A^3}{4} \cos 3t^* \quad . \tag{9.6b}$$

The false resonance comes from the first term on the right-hand side. Since we have not fixed the value of u_1 yet we can simply abolish the troublesome term by equating its coefficient to zero; the resulting equation is called the *modulation equation*; and it yeilds

$$\omega_1 = \frac{3A^2}{8} \quad . \tag{9.7a}$$

the term $\epsilon \omega_1$ in the expansion of w has now been found and u_1 is now free of resonance terms:

$$u_1 = \frac{A}{32} \cos 3t^* . \tag{9.7b}$$

Method of Averaging. For our example we may use an equivalent method. We introduce polar coordinates, relative to Cartesian coordinates u and $v = \frac{du}{dt}$,†

$$u = r \cos \phi , \, v = -r \sin \phi . \tag{9.8a, b}$$

Then (9.1) can be written

$$\dot{r} = 0 + \epsilon r^2 \left(-\frac{\sin 2\phi}{4} + \frac{\sin 4\phi}{8} \right) , \tag{9.9a}$$

$$\dot{\phi} = 1 + \epsilon r^2 \left(\frac{3}{8} + \frac{\cos 2\phi}{2} + \frac{\cos 4\phi}{8} \right) . \tag{9.9b}$$

---

† These coordinates have the topology of Cartesian coordinate but there is of course no natural Euclidean structure in the phase–plane.

We now introduce distorted polar coordinates R, Φ by

$$r = R + \epsilon R_1(R, \Phi) + \cdots \quad , \tag{9.10a}$$

$$\phi = \Phi + \epsilon \Phi_1(R, \Phi) + \cdots \quad . \tag{9.10b}$$

The resulting equations are neglecting $0(\epsilon^2)$

$$\frac{dR}{dt} = 0 + \epsilon \left[A^3 \left(-\frac{\sin 2\Phi}{4} + \frac{\sin 4\Phi}{8} \right) - \frac{\partial R_1}{\partial \Phi} \right] , \tag{9.11a}$$

$$\frac{d\Phi}{dt} = 1 + \epsilon \left[\frac{3}{8} A^2 + A^2 \left(\frac{\cos 2\Phi}{2} + \frac{\cos 4\Phi}{8} \right) - \frac{\partial \Phi_1}{\partial \Phi} \right] . \tag{9.11b}$$

where A is a constant (see below). The functions R_1 and Φ_1 are as yet undetermined; we choose them so as to simplify (9.11) and hence R and Φ as much as possible. To do this we equate $\frac{\partial R_1}{\partial \Phi}$ and $\frac{\partial R_1}{\partial \Phi}$ to the corresponding oscillatory parts of the ϵ-terms in the corresponding equations. By solving the resulting *modulation equations* using a convenient constant of integration we obtain

$$R_1 = -A^3 \left(-\frac{\cos 2\Phi}{8} + \frac{\cos 4\Phi}{32} \right) , \tag{9.12a}$$

$$\Phi_1 = -A^2 \left(\frac{\sin 2\Phi}{4} + \frac{\sin 4\Phi}{32} \right) . \tag{9.12b}$$

and

$$R = A + 0(\epsilon^1) , \tag{9.13a}$$

$$\Phi = (1 + \epsilon \frac{3}{8} A^2)t . \tag{9.13b}$$

Comparison of the Two Methods. Lindstedt's method keeps the original dependent variable u but modifies the time-variable, that is the dependent variable. The modification is simple, the new variable is $t^* = wt$ where w is determined to any desired order in ϵ. In the phase plane where we use $\frac{du}{dt^*}$ as the velocity, orbit is a circle which is modified by higher harmonics of order ϵ, ϵ^2 etc. From a practical point of view, the result is easy to read off.

The method of averaging modifies the dependent variable, r and ϕ instead. While it has many practical virtures, it is also of theoretical interest. We shall only mention one aspect of this: In the phase-plane the solutions of (9.1) are nested closed curves with the origin as a center. Clearly the curves are diffeormorphic to concentric circles. Furthermore, we may deform ϕ on each circle so that u and v are simple harmonic functions of Φ. We may introduce t to form the extended phase-space (sometimes called Cartan space) and we may modify t on each orbit so that the period always becomes 2π. Thus the phase-portrait is diffeomorphic to that of a linear equation, that is the original equation with $\epsilon = 0$. We can see from (9.13) that in using the method of averaging one may construct the linearizing coordinates to any desired order of ϵ.

Nonperiodic Problems. Method of Multiple Scales. Astronomers had used various versions of the method of averaging for many years and those dealing with oscillatory problems in engineering made important contributions, to problems extending far beyond that of the simple periodic oscillator discussed above. An important survey of methods and results was given by Krylov and Boguliubov in a book whose translation by S. Lefschetz appeared in (1947). A decade or so later another method appeared which may be regarded as a generalization of Lindstedt's method.

Consider the damped harmonic oscillator with the equation $u^2 + 2\epsilon \dot{u} + u = 0; 0 < \epsilon \prec 1$. Solutions will contain terms of the type $e^{-\epsilon t} \cos(1-\epsilon^2)^{\frac{1}{2}} t$. This involves two-time scales $\tilde{t} = \epsilon t$ and $t^* = (1-\epsilon^2)^{\frac{1}{2}} t$. The first time-variable resembles the one used in Lindstedt's method, the other one is of a different order, $\tilde{t}/t^* = 0(\epsilon)$. In the late 1950's J.D. Cole introduced the idea of introducing the two scales simultaneous and to determine them by methods generalizing those used in Lindstedt's method.† The

---

† The method became known as the method of multiple scale. This name is unfortunate since the method of matched asymptotic expansions also uses multiple scales. The difference is that in that method each scale is effectively combined to a region (say inner or outer), which is not the case for the damped oscillator.

method was worked out for many examples in J. Kevorkian's Ph.D. thesis (1961) written under Cole's guidance, and in various further publications, e.g. Kevorkian (1962). It was accepted very fast, partly for the negative reason that the work by Krylov, Boguliubov and other Soviet mathematicians was not well understood, and partly because the new method was relatively easy to apply and gave transparent results.†

Recent Theoretical Work. The book by Kevorkian and Cole gives very many examples to illustrate the method. In the meantime the method of averaging has gained wider acceptance. However, there have been other developments. A large number of important relevant theoretical discoveries have been made: the KAM-theory (Kolmogorov-Arnold-Moser), the discovery of strange attractors, chaos, etc. This subject is in a state of rapid development, and time is not yet ripe for discussing it in the style of this book.

---

† There were some isolated instances of independent discovery of similar methods; see, for instance, Mahoney (1962). An important contribution had been made by Kuzman (1959); see the books by Cole and Kevorkian-Cole. However, the vast number of applications in the U.S. stemmed almost exclusively from the orginal work by Cole and Kevorkian.

CHAPTER III
Layer-type Problems.
Partial Differential Equations[†]

3.1. Introduction

We have seen in Chapter II that even restricting ourselves to a few relatively simple-looking ordinary differential equations we get a great variety of types of expansions when applying singular perturbation techniques. Also, as seen in Chapter II, Section 4, replacing the middle term in $\epsilon u'' + u_x - u = 0$ by the quasi-linear term $u u_x$ greatly increases the variety of solutions for two-point boundary value problems. Obviously, we expect the variety of solutions and the techniques necessary to be very large when we consider partial differential equations.

We therefore start by studying analogues of the simplest equation discussed in Chapter I, namely Friedrichs' model equation (I.1.1) and consider first homogeneous equations in two variables of the form

$$ a \frac{\partial u}{\partial x} + b \frac{\partial u}{\partial y} = \epsilon L_2 u \,, \tag{1.1} $$

where a, b are constants and the operator L_2 is a sum of second-order derivatives multiplied by constant coefficients. From the theory of partial differential equations

---

[†] In this chapter we shall use some terminology and theorems which are standard in the theory of partial differential equations. Since there are many textbooks on this subject no particular reference is given.

we know that there are three basic types of such operators which, with a simple choice of coefficients, are

$$L_2 = \frac{\partial^2}{\partial x^2} + \frac{\partial^2}{\partial y^2} = \nabla^2 \quad \text{(Laplace's operator)} , \tag{1.2a}$$

$$L_2 = \frac{\partial^2}{\partial x^2} - \frac{\partial^2}{\partial y^2} \quad \text{(wave operator)} , \tag{1.2b}$$

$$L_2 = \frac{\partial^2}{\partial x^2} . \tag{1.2c}$$

The use of the first of those operators makes (1.1) elliptic, the second makes it hyperbolic and the third makes it parabolic. The qualitative behavior of the solutions and well-posed boundary (or initial) conditions are radically different for the three types.

The reduced equation $(\epsilon = 0)$ of (1.1) is a simple first-order equation, the same for all types. But since the solutions of the full equations are qualitatively different for the three cases we expect the perturbation techniques to vary from case to case. We also observe that another possible generalization is to replace ϵL_2 by $\frac{\partial^2}{\partial x^2} \pm \epsilon \frac{\partial^2}{\partial y^2}$. The reduced equation is then parabolic. We shall call these two cases the elliptic-to-parabolic case and the hyperbolic-to-parabolic case.

In all examples the reduced equations have real characteristics. In Friedrichs' example (and many other examples discussed in Chapter I) the reduced equation had a discontinuity (or several discontinuities). Scaled variables were used to replace the discontinuities by a smooth thin layer. In two variables the reduced problems are expected to have discontinuities across lines. In Lagerstrom-Cole-Trilling (1949) the characteristics of the reduced equation were named subcharacteristics for some problems in fluid dynamics. The reason for a special name was that it was found that the appropriate scaling depended on whether the discontinuity in the reduced equation occurred along a characteristic or not. The special role of subcharacteristics is implicit in the work by Levinson (1950), was emphasized by Latta (1951), illustrated by analysis of exact solutions in Lagerstrom (1964), and several significant examples were given by Cole (1968). Recently a great deal of advanced work on the problem of subcharacteristics has been done by a school of Dutch mathematicians (references are given in Section 2).

While the notion of subcharacteristics (or subcharacteristic direction) has no meaning for ordinary differential equations it is easily generalized to partial differential equations with more than two independent variables and to higher-order equations.

The concept of a *relevant* solution of the reduced equation is of greater importance in partial differential equations than in ordinary differential equations. A relevant solution of the reduced equations is the (outer) limit of some solution of the full equation. Obviously if the full equation represents a physical theory more accurate than the reduced equation only the relevant solutions of the reduced equation make sense as an approximate description of reality. An instructive illustration of the concept of relevant solution is given in Example A4 of Section 2. This example is actually a model example for a problem in the theory of flow with small viscosity. Such a flow is governed by the Navier-Stokes equation; when viscosity is put equal to zero one obtains the Euler equations. The latter equations have a variety of solutions and most of them are not relevant. A major difficulty in the asymptotic theory for viscous flow is that of finding the appropriate relevant solution (see Lagerstrom 1975).

Obviously, it is impossible to give a complete survey of results for partial differential equations in this book. We therefore restrict ourselves to simple illustrative examples of the various phenomena which may occur in studying simple elliptic, hyperbolic and parabolic equations of second order (Sections 2, 3, and 4 respectively). In Section 5 we deal with the equations of incompressible flow as an example of higher-order nonlinear equations. This subject is treated by Van Dyke (1964, 1975); our examples are intended to complement his treatment; we also try to show the importance of dimensional analysis.

3.2. Elliptic Equations

Levinson (1950) made a systematic study of a class of singular-perturbation problems for linear second-order elliptic equations in two variables. He considered the
interior Dirichlet problem for a finite domain, with certain smoothness conditions
on the curve bounding the domain and on the boundary values prescribed there.
The reduced equation was of first order. The subcharacteristics were assumed to
go from one point of the boundary to another (no closed subcharacteristics) and
the boundary had subcharacteristic direction only at isolated points. For a survey
of more general linear elliptic equations, see Eckhaus (1972). Other references to
studies of elliptic equations by the Dutch mathematical school are Grasman (1971),
articles by Grasman and by van Harten in de Jaeger (1973), Grasman (1974) and
for nonlinear equations, van Harten (1975).

In this section we study simple concrete problems; however only the first one
(Example A1) will be of the type discussed by Levinson. Dirichlet's interior problem
is discussed in Subsection A, the exterior problem in Subsection B, assuming in
both cases that the reduced equation is of first order. The elliptic-parabolic case is
discussed in Subsection C. In all cases we assume the full equation to be of second
order. Higher-order elliptic equations are discussed in the section on problems
arising in fluid dynamics (Section 5).

Introductory discussion of a simple equation. In many of the specific
boundary-value problems to be discussed in this section the governing equation
is the very simple one

$$u_x = \epsilon \left(u_{xx} + u_{yy} \right) , \qquad \epsilon > 0 . \tag{2.1}$$

We shall now discuss some properties of this equation, valid for an arbitrary positive value of ϵ, these ideas and results obtained will be useful later on when ϵ is assumed to be near zero or infinity. With appropriate, often obvious changes the discussion will apply to equations similar to (2.1).

Heuristic arguments will play an important role in our discussion of various specific examples. We shall therefore give an intuitive interpretation of (2.1). The independent variables will be interpreted as rectangular coordinates in the plane, the dependent variable will be referred to as temperature and the parameter ϵ as diffusivity or conductivity. This interpretation is of course highly arbitrary. The reader may substitute some other diffusive quantity for temperature, in particular use a probabilistic interpretation of (2.1). In a given physical problem a proper dimensional analysis must be made. We shall assume that this has already been done and that the variables and the parameter in (2.1), as well as in whatever boundary conditions we use, are actually nondimensional. Because of this we do not have to distinguish carefully between temperature and heat (per unit mass or unit volume). Actually, dimensional analysis can be very important in studying perturbation problems as was illustrated in Chapter I Section 3, and will again be stressed in Section 5 below.

Equation (2.1) will be thought of as governing the stationary heat distribution in a fluid whose flow velocity at any point is $\mathbf{i} = (1,0)$. The vector $(u,0)$ describes the transport of heat by the fluid and the vector $-\epsilon(u_x, u_y)$ the conduction (diffusion) of heat. (Here the assumption $\epsilon > 0$ is essential). Equation (2.1) states that the divergence of the sum of the two vectors is zero. It is thus a conservation law; by Gauss' theorem it states that the total heat transported into and conducted into a region is zero. More generally the integral over a closed region of the divergence of the sum of these two vectors equals the total heat sources contained in this region. In an open domain where (2.1) is valid there are no heat sources. However, heat sources may occur at a boundary when the temperature there is prescribed by a boundary condition.

For reasons which will be evident later we shall need the temperature field in an unbounded region due to a stationary heat source of unit strength concentrated at a point. This field is described by the fundamental solution $u = F(x, y; \epsilon)$. (Since the equation is invariant under a translation of the coordinate system we need only consider the case of a source located at the origin.)

We first observe that a general solution (2.1) which vanishes at infinity is obtained by the standard method of making the equation radially symmetric by eliminating the first derivative and then separating variables. Putting

$$u = e^{\frac{x}{2\epsilon}} w(x, y; \epsilon) , \qquad (2.2a)$$

we find

$$w = 4\epsilon^2 (w_{xx} + w_{yy}) . \qquad (2.2b)$$

The solution is then

$$u = e^{\xi} \left[a_0(\epsilon) K_0(\rho) + \sum_{n=1}^{\infty} \left(a_n(\epsilon) \cos n\theta + b_n(\epsilon) \sin n\theta \right) K_n(\rho) \right] , \qquad (2.3)$$

where $2\epsilon\xi = x$, $2\epsilon\eta = y$ and ρ, θ are polar coordinates with respect to ξ and η. The singularity of the fundamental solution $F(x, y; \epsilon)$ at the origin must be the same as that of Laplace's equation, i.e., logarithmic. Hence it is obtained from (2.3) by putting $a_n = b_n = 0$, $n > 0$ and choosing a_0 so that the total heat transported and conducted through a curve enclosing the origin is unity. By choosing that curve to be a circle with center at the origin and ρ very small, one finds the answer

$$F(x, y; \epsilon) = \frac{1}{2\pi} e^{\frac{x}{2\epsilon}} K_0 \left(\frac{r}{2\epsilon} \right) . \qquad (2.4)$$

The leading terms in the expansion of $K_0(z)$ for small and large values of z are, respectively,

$$K_0(z) = -\log \frac{\gamma_0 z}{2} + O(z^2 \log z) , \quad \gamma_0 = e^{\gamma} \doteq 1.781,$$

$$\gamma = \text{Euler's constant}, \qquad\qquad (2.5a)$$

$$K_0(z) \simeq \sqrt{\frac{\pi}{2z}}\, e^{-z}\left(1 - \frac{1}{8z} + \dots\right). \qquad\qquad (2.5b)$$

In the discussion above the only restriction on ϵ was that it be positive. From now on we shall also assume $\epsilon \prec 1$ except in Example B2 where $\epsilon^{-1} \prec 1$.

A. Interior Problems

Example A1. Subcharacteristics crossing the domain. We seek the solution of (2.1) in the interior of a compact simply connected convex domain on the boundary of which the value of u is prescribed (Dirichlet's interior problem). The reduced equation is

$$u_x = 0 , \qquad\qquad (2.6)$$

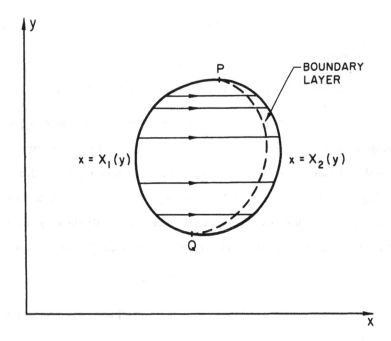

Figure 2.1. Example A2.1.

and hence the subcharacteristics are the lines $y = $ constant and go from one point of the boundary to another. It is specifically assumed that the boundary is differentiable and has subcharacteristic direction (that is, a horizontal tangent) only at two isolated points P and Q. Thus the boundary is divided into two pieces, $x = X_1(y)$ and $x = X_2(y)$, to the left and right of these points as shown in Figure 2.1. We assume the boundary values to be continuous and denote the prescribed value of u along $X_j(y)$ by $U_j(y)$. According to (2.6) the outer solution is constant along subcharacteristics except for possible discontinuities. As in Chapter I Section 1 the location of a discontinuity is determined by the requirement that the discontinuity can be replaced by a local layer which can be matched to the outer solution. As is easily seen, this excludes interior discontinuities. Furthermore, a boundary-layer equation at the left endpoint would give an exponentially growing solution whereas at the right endpoint one can find a solution which decays exponentially from the prescribed value at the boundary to the value of the outer solution.

Note that the flow lines are determined by (2.6) but that this equation does not tell us if the flow goes to the left or to the right. By considering the full equation (with $\epsilon > 0$) we get the direction assumed in the introductory discussion at the beginning of this section. The flow leaves the left endpoint with the temperature prescribed there, namely $u = U_1(y)$ and moves to the right with almost unchanged temperature. Only upon approaching the right endpoint is there a very rapid adjustment to the value prescribed there. Thus, the influence of the temperature prescribed on the left boundary spreads upstream along streamlines but decays exponentially at the right boundary. Heat conduction takes place upstream along the flow lines; to the order considered there is no conduction across (transversal to) the flow lines.

To describe this phenomenon quantitatively we introduce an inner variable $\tilde{x}$

by

$$\epsilon \tilde{x} = X_2(y) - x . \tag{2.7}$$

Then (2.1) may be written

$$\left[1 + \left(X_2'(y)\right)^2\right] \frac{\partial^2 u}{\partial \tilde{x}^2} + \frac{\partial u}{\partial \tilde{x}} + \epsilon \left[2X_2'(y)\frac{\partial^2 u}{\partial \tilde{x}\partial y} + X_2(y)\frac{\partial u}{\partial \tilde{x}}\right] + \epsilon^2 \frac{\partial^2 u}{\partial y^2} = 0 . \tag{2.8}$$

The leading term of the inner expansion then satisfies the boundary-layer equation

$$\left[1 + \left(X_2'(y)\right)^2\right] \frac{\partial^2 g_0}{\partial \tilde{x}^2} + \frac{\partial g_0}{\partial \tilde{x}} = 0 . \tag{2.9}$$

The solution of (2.9) which satisfies the boundary condition $u = U_2(y)$ and matches to the outer solution is

$$g_0(\tilde{x}, y) = U_1(y) + \left[U_2(y) - U_1(y)\right] \exp\left\{-\tilde{x} \cdot \left[1 + \left(X_2'(y)\right)^2\right]^{-1}\right\} . \tag{2.10}$$

This is also the leading term in the composite expansion.

Consider now a subcharacteristic $y = y_0$ in a neighborhood of P (or Q) of order ϵ. Since $X_2(y_0) - X_1(y_0) = O(\epsilon)$ it does not make sense to talk about an exponential decay along this subcharacteristic. On the other hand, if we assume the boundary values to be differentiable, then $U_2(y_0) - U_1(y_0) = O(\epsilon)$ and to order unity we may neglect the inapplicability of boundary-layer theory near P or Q. However, in going to higher-order approximations the small neighborhoods of P and Q must be given special consideration. It is to be expected that both x and y have to be scaled by ϵ so that the full equation is valid in those neighborhoods, although the geometry of the boundary is simplified by an approximation. This point has been examined in detail by Grasman (1971).

Example A2. Segment of the boundary is a subcharacteristic. Let (2.1) again be the governing equation but on the domain shown in Figure 2.2. Its boundary has a segment Q_1Q_2 which lies on the subcharacteristic $y = 0$. Assume the boundary values to be $u = U_1(y)$ on Q_1P, $u = U_2(y)$ on PQ_2 and $u = U_3(x)$ on Q_1Q_2.

Using the boundary values $U_1(y)$ and $U_2(y)$ we construct an asymptotic solution as in the previous example. We subtract this solution from the full solution; thus, to the approximation considered, we might as well have assumed that $U_1(y) = U_2(y) = 0$.

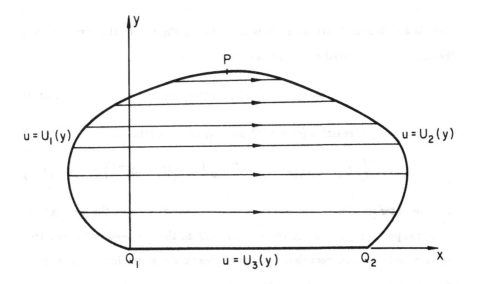

Figure 2.2. Example A2.2.

However, the boundary values along $Q_1 Q_2$ have not yet been taken into account, and because of this a second boundary layer must be inserted. Since the segment $Q_1 Q_2$ is a flow line, this layer cannot be of the type characterized by upstream diffusion. Instead, diffusion occurs transverse to the streamlines, and so the term u_{yy} must appear in the boundary layer equation. For this reason y is scaled as

$$\tilde{y} = y/\sqrt{\epsilon} . \tag{2.11a}$$

The exact equation is now

$$\frac{\partial u}{\partial x} = \frac{\partial^2 u}{\partial \tilde{y}^2} + \epsilon \frac{\partial^2 u}{\partial x^2} \, , \tag{2.11b}$$

and the leading order boundary-layer correction $h_0(x, \tilde{y})$ satisfies the equation

$$\frac{\partial h_0}{\partial x} = \frac{\partial^2 h_0}{\partial \tilde{y}^2} \, , \tag{2.12a}$$

which is a diffusion (heat) equation with x playing the role of the time variable. The boundary and initial conditions are

$$h_0(x, 0) = U_3(x) \, , \quad h_0(Q_1) = 0 \, . \tag{2.12b}$$

This is a standard radiation (signaling) problem whose solution is

$$h_0(x, \tilde{y}) = \int_0^x U_3(\rho) \tilde{y} [r\pi(x - \rho)]^{-1/2} \exp\left\{ -\tilde{y}^2 [4(x - \rho)]^{-1} \right\} d\rho \, . \tag{2.13}$$

Since the outer solution is zero this boundary-layer solution is also the leading term of the composite expansion. A reasoning similar to the one used in the previous example leads us to suspect that in finding higher order approximation a separate expansion will be needed near the point Q_2.

Example A3. Discontinuous boundary values. Consider again (2.1) on the unit disc centered at the origin with boundary values which are differentiable except for one discontinuity on the left-hand side of the circle. More specifically, assume that the value of u on $r = 1$ changes from $+1$ to -1 as θ increases from $-\pi$ to $+\pi$. The outer solution is then discontinuous across the subcharacteristic $y = 0$. However, solutions of elliptic equations are known to be smooth in the interior of the domain under consideration. Thus the discontinuity must be replaced by a layer. Since the outer solution increases from -1 to $+1$ for fixed x we expect u_{yy} to be more important than u_{xx} in the layer equation which thus must involve a scaling of y. As in Example A2 we find that a suitable inner variable is $\tilde{y} = y\epsilon^{-1/2}$. However, the layer is now a free layer rather than a boundary layer as was the case in the previous example. The function $g_0(x, \tilde{y})$ describing the layer must satisfy the equation

$$\frac{\partial g_0}{\partial x} = \frac{\partial^2 g_0}{\partial \tilde{y}^2} , \qquad (2.14a)$$

the matching conditions

$$g_0(x, -\infty) = -1 , \quad g_0(x, \infty) = 1 , \qquad (2.14b, c)$$

and the initial condition

$$g_0(-1, \tilde{y}) = \begin{cases} +1, & \text{if } \tilde{y} > 0, \\ -1, & \text{if } \tilde{y} < 0. \end{cases} \qquad (2.14d)$$

Thus

$$g_0(x, \tilde{y}) = \operatorname{erf} \tilde{y}[4(x+1)]^{-1/2} . \qquad (2.15)$$

Example A4. Closed subcharacteristic. A function $u(r, \theta)$ is defined for $r \le 1$ by the equation

$$\frac{\partial u}{\partial \theta} = \epsilon \left(\frac{\partial^2 u}{\partial r^2} + \frac{1}{r} \frac{\partial u}{\partial r} + \frac{1}{r^2} \frac{\partial^2 u}{\partial \theta^2} \right) , \qquad (2.16a)$$

and the boundary condition

$$u(1, \theta) = U(\theta) . \qquad (2.16b)$$

The subcharacteristics are thus the circles $r = $ constant and the leading term of the outer solution is a function of r only

$$u \to f_0 = F(r) . \qquad (2.17)$$

We are now faced with a new situation: Since the subcharacteristics are closed there are no obvious boundary conditions to determine $F(r)$. To determine $F(r)$ we integrate (2.16a) over each disc $0 < r \le$ constant ≤ 1. The left-hand side gives zero and we apply the divergence theorem to the right-hand side. This gives

$$0 = \int_0^{2\pi} \left(\frac{\partial u}{\partial r} \right) r \, d\theta = r \frac{\partial}{\partial r} \int_0^{2\pi} u \, d\theta , \quad \text{at any } r = \text{constant} . \qquad (2.18)$$

Note that this result is independent of ϵ. As ϵ tends to zero the outer limit of u becomes independent of θ (Equation 2.16). Combining this with (2.18) we find that the outer limit of u is also independent of r for $r < 1$, hence it is a constant. The value of the constant is found, again from (2.18), to be the mean value

$$u = \overline{U} = \frac{1}{2\pi} \int_0^{2\pi} U(\theta) d\theta \ . \tag{2.19}$$

If $U(\theta)$ is not constant there is thus a discontinuity along the subcharacteristic $r = 1$ which will have to be replaced by a boundary layer of thickness $\sim \epsilon^{1/2}$. We write the exact equation

$$\frac{\partial u}{\partial \theta} = \frac{\partial^2 u}{\partial \tilde{r}^2} - \frac{\epsilon^{1/2}}{(1 - \epsilon^{1/2}\tilde{r})} \frac{\partial u}{\partial \tilde{r}} + \frac{\epsilon}{(1 - \epsilon^{1/2}\tilde{r})^2} \frac{\partial^2 u}{\partial \theta^2} \ , \tag{2.20a}$$

where $\tilde{r}$ is the boundary-layer coordinate

$$\epsilon^{1/2}\tilde{r} = 1 - r \ . \tag{2.20b}$$

Thus the boundary-layer equation for $g_0(\tilde{r}, \theta)$ is

$$\frac{\partial g_0}{\partial \theta} = \frac{\partial^2 g_0}{\partial \tilde{r}^2} \ , \tag{2.21a}$$

with the boundary conditions

$$g_0(0, \theta) = U(\theta) \ , \quad g_0(\infty, \theta) = \overline{U} \ . \tag{2.21b, c}$$

While (2.21a) looks like the heat equation one must remember that θ is an angular variable. The customary initial condition is thus replaced by the requirement that g_0 be 2π-periodic in θ. The problem can now be solved by standard methods of separation of variables and Fourier analysis. The details are left to the reader (see Exercise 2.1).

The new features of the present example are due mainly to the fact that the subcharacteristics are closed. It may therefore be of interest to look at the solution from a heuristic point of view. Without writing down the equation explicitly we consider time-dependent heat conduction in the domain $r \leq 1$. The boundary conditions are given by (2.16b) and the initial conditions are arbitrary. As time tends to infinity the temperature field tends to a steady-state limit which is the solution of (2.16a, b). The crucial fact is that the streamlines are closed so that the heat conductivity has infinite time to act on the fluid which is constrained to a finite domain. (In previous problems fluid particles entered through one part of the boundary and then disappeared through the other part.) Thus, heat conduction acting on the same fluid over an infinite time will tend to even out temperature differences within the fluid. If the prescribed value at the boundary were a constant the value of u at every interior point would tend to that constant. If $U(\theta)$ is not constant its variation will offset the uniformising influence of heat conduction in the interior. As a matter of fact, the *smaller* ϵ is the greater the uniformity of temperature in the interior in the sense that u will be essentially constant since the influence of the boundary is negligible except in a thin layer of thickness $\epsilon^{1/2}$. The fluid particles, which move in circular paths, are alternately heated and cooled by the boundary but this effect becomes negligible outside the boundary layer. Thus the essential feature of the present example, both for the heuristic reasoning and the analytical approximation, is the fact that the flow lines are closed.[†]

B. Exterior Problems

The equations to be discussed below will be the same as those in Subsection A but we shall now study asymptotic solutions of the *exterior* Dirichlet problem. The values of the unknown function will be prescribed on the boundary of a simply connected finite or semi-infinite domain. We seek the solution in the infinite region *outside* the

---

[†] The reasoning above (analytical or heuristic) shows that the only *relevant* solution of the reduced equation is $f_0 = $ constant. The actual value of the constant depends on the boundary condition (2.16b).

domain. To ensure uniqueness of the solution a boundary condition at infinity is
needed; we shall encounter the same phenomena as in Subsection A and in addition,
some new phenomena. One is the appearance of an infinite downstream wake as in
Example B1. A second one is the possible occurrence of an artificial parameter as in
Example B2. Finally, Example B3 shows that a singular perturbation problem may
arise even when the small parameter multiplies only the lowest-order derivatives.

Example B1. Flow past a circular cylinder. Diffusivity small. We consider again
(2.1) for small positive values of ϵ but seek the solution *outside* a unit circle (cylinder) with center at the origin. The boundary conditions are

$$u = 1 \text{ at } r = 1 ; \quad u = 0 \text{ at } r = \infty . \qquad (2.22a, b)$$

Strictly speaking, the flow is *through* rather than *past* the circle since the flow lines
(subcharacteristics) are the lines $y = $ constant which pass straight through the
circle. For $x \leq 0$ an approximate solution is easily constructed using the same ideas
as in Example A1. Cold $(u = 0)$ flow arrives from upstream infinity $(x = -\infty)$.
To the first approximation the flow is unaffected by the cylinder for $x \leq 0$ and
$|y| \geq 1$. For $|y| \leq 1$ the higher $(u = 1)$ temperature spreads upstream against the
flow but its influence decays exponentially. Denoting the front half of the cylinder
by

$$x = X_2(y) \equiv -\sqrt{1 - y^2} , \qquad (2.23)$$

and defining an inner variable $\tilde{x}$ by

$$\epsilon \tilde{x} = X_2(y) - x , \qquad (2.24)$$

we find, in complete analogy with (2.6), the boundary-layer solution for $x \leq 0$,

$$g_0 = \exp\left[-\tilde{x}\left(1 + X_2'(y)\right)^{-2}\right] , \quad \text{for } x \leq X_2(y), \ |y| < 1 , \qquad (2.25a)$$

$$g_0 = 0 , \quad \text{for } x \leq 0 , \ |y| > 1 . \qquad (2.25b)$$

This solution is obviously uniformly valid in the region indicated since it contains the outer solution which is identically zero.

In order to find the solution for $x \geq 0$ it is instructive to consider the solution inside the cylinder, although it is not part of the solution of the original boundary-value problem. Obviously, the constant function $u = 1$ satisfies both the boundary conditions and the equations inside the cylinder. Thus, as a streamline enters the cylinder it is not analytic at the front half of the boundary. This is true both for the exact solution and the asymptotic approximation which we are trying to construct. For future reference we note that there must then be heat sources distributed along $x = X_2(y)$; their total strength is $= 2$ in the boundary-layer approximation as is easily seen by comparing the flow of heat through $x = -\infty$, $|y| \leq 1$ and $x = 0$, $|y| \leq 1$.

At the rear half of the circle,

$$x = X_1(y) \equiv \sqrt{1 - y^2} , \quad |y| \leq 1 , \qquad (2.26)$$

there is no boundary layer on the left-hand side $\left(x \leq X_1(y)\right)$ since the arriving flow has the required value $u = 1$. This is of less immediate interest here. However, disregarding the interior flow, we see that there is also no boundary layer for $x \geq X_1(y)$. This follows from the general principles used in discussing Example A1. We conclude that the outer solution for $x \geq X_1(y)$, $|y| < 1$ is $f_0 = 1$. On the other hand the correct boundary value at downstream infinity $(x = +\infty)$ is $u = 0$ so that the outer solution is discontinuous there. This might seem to necessitate a boundary-layer at infinity, but we do not have to worry about how to construct such a layer. The outer flow is zero for $|y| > 1$ (any value of x). Thus there are discontinuities along the two characteristic half-lines, $x > 0$, $y = \pm 1$. We know from Example A3 that these discontinuities must be replaced by free layers across which transversal diffusion takes place. The equation for these layers are obtained by dropping the term ϵu_{xx} in (2.1) and the solutions are (*cf.* Equation (2.13))

$$g_u = \tfrac{1}{2} - \tfrac{1}{2} \, \text{erf} \, \frac{\tilde{y}_u}{\sqrt{4x}} \, , \quad \epsilon^{1/2} \tilde{y}_u = y - 1 \, , \qquad (2.27a)$$

for the upper layer (around $y = +1$) and

$$g_\ell = \tfrac{1}{2} - \tfrac{1}{2} \, \text{erf} \, \frac{\tilde{y}_\ell}{\sqrt{4x}} \, , \quad \epsilon^{1/2} \tilde{y}_\ell = y + 1 \, , \qquad (2.27b)$$

for the lower layer around $y = -1$.

The field behind the circle where u is not negligibly small will be called the *wake*. For $x > 0$ the value of u in each layer is greater than zero for any value of y. However, each layer has an effective width outside of which u is negligible.[†] Previously, we have described free layer with transversal diffusion as being of thickness, or width, $\sim \sqrt{\epsilon}$. Note however that, for instance, the upper layer is a similarity solution of the heat equation which depends on the similarity variable $(y - 1)(\epsilon x)^{-1/2}$. Since x grows indefinitely x-dependence of the width cannot be ignored. Thus the width of each layer is more properly described as $\sim \sqrt{\epsilon x}$, and it thus increases parabolically downstream. Even though the constant of proportionality for the width is left unspecified, it is clear that for x sufficiently large, that is of order ϵ^{-1}, the two layers merge. Since we are dealing with a linear equation the description of the merged layers causes no difficulty; one simply adds the functions g_u and g_ℓ of (2.27).

Thus far downstream the wake increases parabolically as $\sim \sqrt{\epsilon x}$. It is important to derive its shape by a method which is valid even if there is nonlinear interaction between the two free layers described above. Such a method exists if there is an appropriate conservation law, as is the case for our present problem. We have seen that the total strength of the heat sources produced by the cylinder is $= 2$. In studying the field far from the cylinder we may assume the heat sources

---

[†] Obviously, the extent of the wake region is defined imprecisely.

to be concentrated at the origin. The field is thus described by $2F(x, y; \epsilon)$ where the fundamental solution F is given by (2.4). However, since we study the flow field at large distances from the origin we may as well evaluate the K_0-function for large values of its argument, using (2.5b). This gives

$$F(x, y; \epsilon) \simeq \sqrt{\frac{\epsilon}{4\pi r}} e^{\frac{x-r}{2\epsilon}} . \qquad (2.28a)$$

Thus the effect of a heat source decreases exponentially upstream, consistent with what we have found for upstream boundary layer. In fact, the above expression shown exponential decay except when $y \prec x$ in which case we may write it as

$$F \simeq \sqrt{\frac{\epsilon}{4\pi x}} e^{-\frac{y^2}{4\epsilon x}}, \quad y \prec x, \quad 1 \prec x . \qquad (2.28b)$$

On any fixed parabola $y^2 = x \cdot$ constant, including the positive x-axis the decay of u with x is algebraic rather than exponential.

The approximate formula (2.28b) can be obtained directly. In the region studied we expect the y variations to be more rapid than the x-variations. Hence we may neglect u_{xx} relative to u_{yy}. The resulting equation is the heat equation with x playing the role of time. (Note that neglecting u_{xx} implies neglecting any upstream influence.) We have previously encountered this equation as a suitable equation for boundary layers or free layers. These could be matched to some outer flow. *In the present case* (2.28b) *cannot be matched.* The governing equation has not been obtained by a limit process. It is valid for x large but we have no suitable parameter to describe this. The fact that a suitable parameter does not exist will play an important and often overlooked role in the next example and in some examples in Section 5.

Example B2. Flow past a semi-infinite flat plate. Diffusivity small. Consider again Equation (2.1) and let the boundary conditions be

$$u = 1 \text{ as } r \uparrow \infty, \quad u = 0 \text{ for } x \geq 0, \quad y = 0 . \qquad (2.29)$$

Using the same ideas as in previous problems we find that the outer solution is discontinuous along the "plate" $x \geq 0$, $y = 0$. Introducing the inner coordinate $\tilde{y} = y\epsilon^{-1/2}$ we find a local boundary layer where heat is conducted transversally but not upstream. The solution is of similarity type and includes the outer solution. It is

$$u = 0 \text{ for } x < 0, \quad u = \text{erf} \left(\frac{y^2}{4\epsilon x} \right)^{1/2} \quad for \ x > 0 . \qquad (2.30)$$

This solution is not analytic across the plate. The reason for this is obvious since the plate is represented by singularities (stationary heat sources).

Parabolic Coordinates. We now introduce parabolic coordinates ξ, η by

$$x + iy = (\xi + i\eta)^2 , \qquad (2.31)$$

with the restrictions

$$\eta \geq 0 , \quad \xi \text{ and } y \text{ have the same sign} . \qquad (2.31b, c)$$

Some restrictions are necessary since the Riemann surface of the square-root function is two-sheeted. The complex $(x + iy)$-plane is cut along the plate which now is $\eta = 0$ and $\xi > 0$. The function $\xi(x, y)$ is not analytic across the plate. In parabolic coordinates the full equation (2.1) reads

$$\xi u_\xi - \eta u_\eta = \frac{\epsilon}{2} (u_{\xi\xi} + u_{\eta\eta}) . \qquad (2.32)$$

If the boundary-layer analysis is done as before the term $u_{\xi\xi}$ is found to be negligible. The resulting equation as well as the boundary conditions are then invariant under the mapping $\xi \to k \cdot \xi$ and a similarity analysis then indicates that in the boundary layer u is a function of η only. But then $u_{\xi\xi}$ is identically zero and one finds a boundary-layer solution which is also an *exact* solution of the original problem,

$$u(\eta; \epsilon) = \text{erf} \left(\eta \epsilon^{-1/2} \right) . \qquad (2.33)$$

Obviously (2.30) and (2.33) represent different solutions (see Exercise 2.4). If (2.30) is rewritten in parabolic coordinates nothing is changed in the content of that solution. On the other hand, if one first expresses the exact solution in parabolic coordinates and then performs the boundary layer analysis one obtains a different and better result.[†] This phenomenon has been analyzed systematically for boundary-layer theory in fluid mechanics by Kaplun (1954); his results will be discussed briefly in Section 5.

The problem of uniform validity. Instead of pursuing further the question of the role of the coordinate system, we shall ask a somewhat related question. The solution (2.33) is obviously valid to any order, being exact. The solution (2.30) includes both the boundary-layer solution and the outer solution. We would then hope that, without being exact, it is at least uniformly valid to order unity. However, as pointed out by Lagerstrom and Cole (1955; Section 6.1), an approximation which is uniformly valid to order unity *must* be the exact solution. A somewhat strengthened version of this assertion is given by Chang (1961).

We shall discuss the proof only for our special case; this will demonstrate the general idea. The basic feature is that if we introduce $\tilde{y} = y\epsilon^{1/2}$ into the exact equation and boundary conditions then ϵ disappears completely from the problem. If the solution exists and is unique it must then have the form $u = u^*(x, \tilde{y})$. Assume now that w^* is an approximation *uniformly* valid to order unity *which has the same similarity structure as* u^*, that is $u = w^*(x, \tilde{y}) + o(1)$. (Here "similarity" means that y and ϵ can be combined into one variable $\tilde{y}$, not the similarity referring to the combination of coordinates used above.) We can state the assertion more succinctly as follows:

Let $u^*(x, y; \epsilon)$ and $w^*(x, y; \epsilon)$ be functions of the two arguments x and

---

[†] So far, the material for the present example has been taken from Lagerstrom, Cole and Trilling (1949). However, the further discussion in that report is partially superseded.

$ye^{-1/2}$ only. If then $d \equiv |u^* - w^*| = o(1)$ uniformly, it follows that

$w^* = u^*$ everywhere for all positive ϵ. (2.34)

PROOF: By assumption, given any $\delta > 0$ there exists an $\epsilon(\delta) > 0$ such that $d < \delta$ for all x and y and all ϵ such that $0 < \epsilon < \epsilon(\delta)$: We want to show that $d < \delta$, and hence $w^* = u^*$, for all x and y and all positive ϵ. Put $\epsilon(\delta) = \eta$. Then, by similarity, $d(x, y; \epsilon) = d(x, y\sqrt{\eta/\epsilon}; \eta)$ for any $\epsilon > 0$ and any x and y. By assumption the right-hand side of this equality is $< \delta$. thus $d(x, y; \epsilon)$ is arbitrarily small and hence $w^* = u^*$. □

Note that it is important for the proof, and for the validity of the theorem, that w^* have the same similarity as u^*. If one puts $w^* = u^* + \epsilon$ then obviously w^* is an approximation uniformly valid to order unity but it does not have the required similarity. The specific form of the similarity is irrelevant. What matters is that one can eliminate ϵ completely from the problem by absorbing it in one (or both) of the coordinates. Such a parameter which can be eliminated completely will be called an *artificial parameter*. We shall discuss the problem of artificial parameters further in Section 5. There it will be seen that the term "artificial" has a mathematical rather than a physical meaning: Viscosity, which is a physically very respectable parameter, may become mathematically artificial for certain boundary-value problems.

Obviously one may ask the question: If ϵ is artificial, why use it at all? First of all, for the same equation ϵ may be artificial only for certain boundary-value problems. Secondly, even for the present example, the use of ϵ in boundary-layer technique helped us to get some kind of approximate solution. In Lagerstrom and Cole (1955, Section 6) it was pointed out that while normally the use of an artificial parameter gives us a boundary-layer approximation which is not uniformly valid, there are occasionally cases of "exceptional luck" in which we get the exact solution. In the present example we had exceptional luck if we used parabolic coordinates. The exact solution so obtained may then be used to analyze validity of the boundary-layer solution obtained by using rectangular coordinates and the artificial parameter. This is left to the reader (Exercise 2.4). For further discussion of artificial parameters, see Section 5 and references given there.

Example B3. Flow past a circular cylinder. Diffusivity large. Consider the same boundary-value problem as in Example B1 but assume $\epsilon \uparrow \infty$. Putting $\eta = \epsilon^{-1}$ we see that the small parameter η multiplies only the lowest-order derivative, namely u_x. To facilitate comparison with a corresponding problem in Section 5 (low Reynolds number flow) we shall make a change in the boundary conditions: We shall assume $u = 0$ at $r = 1$ and $u = 1$ at $r = \infty$. Since we have a simple linear problem this change is actually trivial both as regards the analysis and the heuristic reasoning. Furthermore, we shall also make a change in notation by writing $\tilde{x}, \tilde{y}, \tilde{r}$ instead of x, y, r. The problem is thus

$$\eta u_{\tilde{x}} = u_{\tilde{x}\tilde{x}} + u_{\tilde{y}\tilde{y}} , \tag{2.35a}$$

$$u = 0 \text{ at } \tilde{r} = 1 ; \quad u = 1 \text{ at } \tilde{r} = \infty . \tag{2.35b, c}$$

If we put $\eta = 0$ $(\epsilon = \infty)$ we obtain Laplace's equation, which has no solution satisfying both boundary conditions. It thus appears that we have a singular perturbation problem.

In terms of the heuristic reasoning used earlier a hot $(u = 1)$ flow from upstream infinity, $x = -\infty$, is cooled by a circular cylinder whose surface temperature is kept at $u = 0$. This temperature is maintained by a distribution of (negative) heat sources on the body surface $(\tilde{r} = 1)$. Putting $\eta = 0$ corresponds to zero flow and, in two dimensions, it is then impossible to have heat sources of total strength different from zero which give a stationary bounded temperature field in the domain $\tilde{r} > 1$.

Using singular perturbation techniques and arguments similar to those used in

Chapter II Section 7, one sees that at any fixed point $(\bar{x}, \bar{y})$ u tends to zero as η tends to zero. On the other hand, putting $x = \bar{x}\eta$, $y = \bar{y}\eta$, $r = \bar{r}\eta$ one finds that at (x,y) fixed u tends to unity as η tends to zero. It is left to the reader to justify this heuristically and to continue the reasoning to find matching inner and outer solutions (Exercise 2.5).

Another way of proceeding is to use the general solution (2.3) in suitable co-ordinates and find a solution which satisfies the boundary conditions to a certain order. Again it is left to the reader (Exercise 2.6) to carry out the details of this method and to compare the results with those obtained by the first method.

The preceding example, as well as the examples in Chapter II Section 5, are model examples for the problem of flow at low Reynolds numbers which will be discussed in detail in Section 5. In Exercise 2.7 we give a further model example for which the basic equation is a nonlinear partial differential equation, but which still is simpler then the real physical problem to be discussed in Section 5.

C. Elliptic-to-Parabolic Case

The basic equation for our examples is

$$u_x = u_{yy} + \epsilon u_{xx} , \quad 0 < \epsilon < 1 . \tag{2.36}$$

This equation was already studied by Latta (1951) who proved a theorem for the interior problem about the structure of an approximation whose difference with the exact solution is uniformly $O(\epsilon)$. Here we shall not discuss the general theorem but shall illustrate the expected behavior of the solutions with the aid of some simple boundary-value problems.

As usual we shall give an intuitive interpretation of (2.36) which is convenient but which may be replaced by other interpretations according to the reader's taste. If we put $\epsilon = 0$, the heat equation results, the variable x corresponding to "time." However, then u_{xx} has no convenient interpretation. We therefore regard (2.36) as an equation for stationary heat distribution in a moving fluid: x and y are rectangular coordinates and u a diffusive quantity ("heat") which is transported with the velocity i and at the same time diffuses. The subcharacteristics are the streamlines $y = $ constant. This resembles the interpretation given to (2.1) except for one major difference: For ϵ small the diffusion is not small but highly anisotropic; for $\epsilon = 0$ the diffusion is entirely transversal and expressed by the term u_{yy}. An essential feature of the problem can already be seen if we impose the conditions

$$u(0,y) = 1, \quad u(1,y) = 0 \ . \qquad\qquad (2.37a,b)$$

There is no transversal diffusion since the temperature is already uniform in the y-direction. The problem reduces to a two-point boundary-value problem for an ordinary differential equation similar to the ones treated in Chapter II Section 1. The temperature remains almost constant $(u \simeq 1)$ along streamlines in the interval $0 \leq x < 1$ except that when the streamline approaches the boundary $x = 1$ the temperature adjusts itself very fast (exponentially) in a layer of thickness ϵ to the required value $u = 0$. (Clearly, this asymptotic solution shows that it is convenient for the intuitive interpretation to regard the flow direction for $\epsilon > 0$ to be the positive x-direction.)

The trivial example just mentioned showed the weak upstream diffusion. The effect of transversal diffusion can be seen from the more general but still very simple example given in Exercise 2.8.

In Examples A2 and A3 discontinuities in the boundary conditions led to discontinuities in the outer solution which were smoothed out by transversal diffusion in a layer of thickness $\sqrt{\epsilon}$. In (2.36), however, transversal diffusion is of order unity and discontinuities in the boundary conditions are smoothed out already in the leading term of the outer solution. This is illustrated by Exercise 2.9.

Exercises for §2

2.1. Solve the boundary-layer equation for Example A4, assuming that $U(\theta)$ of (2.16b) can be developed in a cosine series. Give a composite expansion uniformly valid to order unity.

2.2. By separation of variables construct the solution of (2.16a) when $U(\theta) = \sum_{n=0}^{\infty} a_n \cos n\theta$. Discuss the variation with ϵ for $0 \le \epsilon < \infty$ and study directly the limiting cases $\epsilon \downarrow 0$ and $\epsilon \uparrow \infty$. Justify the asymptotic analysis of Exercise 2.1.

2.3. For the preceding two exercises discuss the consequences of letting ϵ be in the range $-\infty < \epsilon < 0$.

2.4. a) How does the solution of 2.30 behave on and near the line $x = 0$?

b) Express (2.33) in the coordinates x and $\tilde{y} = y\epsilon^{1/2}$. Recover (2.30) by a suitable limit process and discuss the next term obtained by this limit process. Discuss the general nature of the complete series obtained by this process.

2.5. Using arguments similar to those of Chapter II Section 7 find the first two terms of the outer expansion and the first two terms of the outer expansion and the first term of the inner expansion for the problem (2.35), $0 < \eta \prec 1$.

2.6. Using (2.3), in suitable coordinates, find an exact solution of (2.35a) which satisfies (2.35c) exactly and (2.35b) approximately. Compare the results with those of Exercise 2.5 above.

2.7. a) We replace (2.35a) by the nonlinear equation

$$\eta u u_{\tilde{x}} = u_{\tilde{x}\tilde{x}} + u_{\tilde{y}\tilde{y}},\qquad\qquad(2.38)$$

and keep the boundary conditions (2.35b, c). Solve the problem of Exercise 2.5, but with the modified equation (2.38).

b) Same as a) except that we now consider three dimensions. The equation is thus

$$\eta u u_{\tilde{x}} = u_{\tilde{x}\tilde{x}} + u_{\tilde{y}\tilde{y}} + u_{\tilde{z}\tilde{z}},\qquad\qquad(2.39)$$

and $\tilde{r}$ stands for the three-dimensional radius. In addition, show that the leading term of the inner expansion satisfies the boundary condition at infinity, but that the term of $O(\eta)$ does not. Use either Cartesian or cylindrical coordinates.

2.8. Find the first term of the outer, inner, and composite expansion respectively of the solution of (2.36) with the boundary conditions $u(0,y) = \cos ny$, $u(1,0) = 0$.

2.9. Find an approximation uniformly valid to order unity of solutions of (2.36) with boundary conditions

   a) $u(0,y) = 1$ for $y > 0$, $u(0,y) = 0$ for $y < 0$, $u(1,y) = 0$.

   b) $u(0,y) = 1$ for $y > 0$, $u(x,0) = 0$ for $x > 0$, $u(1,0) = 0$.

3.3. Hyperbolic Equations

Introduction. We shall assume that the reader is familiar with the basic classical concepts in the theory of hyperbolic equations such as signal velocity, characteristic directions, propagation of discontinuities, Cauchy's problem, radiation problems, and regions of influence. These are discussed at length and illustrated by many examples in Courant-Hilbert (Vol. II, 1962), in Garabedian (1964) and in many other standard texts.

The hyperbolic nature of an equation and the signal velocity are determined by the highest derivatives alone. Thus the effect of a disturbance (say, of initial conditions or part thereof) is strictly zero outside the region of influence of the disturbance. It may, however, occur that the bulk of the effect of the disturbance, or some special property, is propagated with a velocity less than that of the signal velocity.

Whitham (1974) has given a thorough discussion of various physical examples of waves, using a more intuitive and less formal classification of partial differential equations than the classical one. His treatment of wave hierarchies, group velocity, the Burgers equation, and the Korteweg-de Vries equation illustrates important perturbation methods. However, here we discuss only very simple problems using methods employed elsewhere in this book. Furthermore, we shall not discuss real physical problems directly and we assume that the variables and parameters used have already been made nondimensional.

We shall give some simple examples of the importance and the nonimportance of the signal velocity and wave front. Consider first the "pure" wave equation

$$\Box\, u = 0 \, . \tag{3.1}$$

Here

$$\Box \;=\; \text{wave operator} \;=\; \frac{1}{c^2} \frac{\partial^2}{\partial t^2} - \nabla^2 \, ,$$

and ∇^2 is the Laplacian in n spatial variables (we shall mainly consider the case $n = 1$, so that $\nabla^2 = \frac{\partial^2}{\partial x^2}$). The parameter c is the constant *signal* velocity, a concept which must be carefully distinguished from that of phase velocity and group velocity.[†]

For the solution of the "pure" wave equation

$$\Box\, u = 0 \, , \tag{3.2}$$

the wave fronts and hence the signal velocity play a dominant role. In particular the wave front carries the bulk of the signal sent out from a point (see the discussion of Huygens principle in classical texts). Furthermore, in this case the signal velocity is the only speed which we need consider in Cauchy's problem.

[†] The distinction between the terms "speed" and "velocity" is usually not observed in this context. No confusion need arise since the direction of propagation of a disturbance is given when needed.

Let us now consider the opposite extreme. In the limit of the signal velocity tending to infinity a hyperbolic equation turns into a parabolic equation. Examples are the Burgers equation

$$u_t + u u_x = u_{xx} \, , \tag{3.3}$$

and the Korteweg-de Vries equation

$$u_t + u u_x = u_{xxx} \, . \tag{3.4}$$

From the fact that the signal velocity is infinite we deduce a correct but almost useless piece of information: If initial conditions are given at $t = 0$ on a finite region of the x-axis, then for any $t > 0$, no matter how small, and for any $|x|$, no matter how large, there is some effect of the initial conditions. However, it dies off rapidly, say exponentially, with $|x|$. Clearly, the bulk of the signal is transmitted with a finite velocity determined by the lower-order derivatives. Equation (3.3) will be briefly discussed in Section 4. A thorough discussion of both equations is given by Whitham (*loc. cit.*).

Second-order equations in two variables. Normal forms. We shall assume that the highest-order terms are linear (possibly with variable coefficients) and that the equation is hyperbolic in the entire (x, t)-plane (the characteristics are real and the angles between them at any point are bounded from below by a positive constant). As coordinates we may then introduce characteristic coordinates and, possibly, new (x, t)-coordinates related to the characteristic coordinates by

$$r = t - x, \quad s = t + x \, . \tag{3.5}$$

Even if the (x, t) differ from the original ones we still refer to the new x and t as spatial coordinate and time coordinate respectively.

By adjusting the length and time scales we may make $c = 1$. The wave operator is then

$$\Box = \frac{\partial^2}{\partial t^2} - \frac{\partial^2}{\partial x^2} = 4 \frac{\partial^2}{\partial r \partial s} \, . \tag{3.6}$$

The equations to be studied have the general form

$$N(x, t, u, u_t, u_x) + \epsilon \, \Box^2 u = 0 \ , \tag{3.7}$$

where, as usual, we assume $0 < \epsilon \prec 1$. We want to investigate whether singular perturbation techniques, in particular matched asymptotic expansions, may be applied to (3.7).

Cole (1968) made a pioneering study of this problem (see also Kevorkian-Cole, 1981). He restricted himself to the case of linear equations with constant coefficients, and to the initial-value problem (Cauchy's problem) and the initial-boundary value problem (signal or radiation problem). He made essential use of the idea of sub-characteristics. This work was continued by de Jager (1975) who considered linear equations with variable coefficients and provided proofs and order-estimates of errors, and by Geel (1978) who also considered equations for which the operator N of (3.7) was quasi-linear, and studied the characteristic boundary value problem for linear equations. Geel's book contains an extensive bibliography. Thus de Jager and Geel extended Cole's work, while retaining his basic ideas, and provided proofs and *a priori* estimates. However, Cole had omitted any discussion of some basic restrictions on his method (see end of this section), and subsequent work by de Jager and Geel do not discuss these problems either.

First we shall make some general observations, some of which are useful formulas or theorems while others are conventions.

For convenience, initial values will be given on the x-axis and boundary values (if any) on the positive t-axis.

We always look for solutions for $t > 0$. The direction of the characteristics and subcharacteristics is that of increasing t. Note that the value of the characteristic coordinates given by (3.5) increase with t for fixed x.

Any acceptable method must work also for the case when the initial data are zero outside a finite interval of the x-axis. This gives an important test of any perturbation method: The solution has to vanish outside the region of influence (although, since the solution is only asymptotic, we may have to allow it to be transcendentally small there).

In dealing with hyperbolic equations one often considers problems with discontinuities in the initial (or boundary) conditions. It is typical that these discontinuities are not smoothed out, but propagated along the characteristics, and only along the characteristics. Problems with discontinuities in the given conditions are typical for hyperbolic equations and any acceptable perturbation method must be able to cope with such problems.

There is a final restriction on the solution which is physical rather than mathematical. It is, of course, possible to find values of the constants in (3.7) so that mathematically exact solutions grow with time, for instance exponentially. However, we shall assume restrictions on the constants so that such solutions are not possible.

Linear equations with constant coefficients. We consider the following special case of (3.7),

$$Au + Bu_t + Cu_x + \epsilon \left(\frac{\partial^2 u}{\partial t^2} - \frac{\partial^2 u}{\partial x^2} \right) = 0 . \tag{3.8}$$

If we put

$$u = e^{-\alpha t} v , \tag{3.9a}$$

then

$$Au + Bu_t + \epsilon u_{tt} = \left[(A - \alpha B + \epsilon \alpha^2)v + (B - 2\alpha\epsilon)v_t + \epsilon v_{tt} \right] e^{-\alpha t} , \tag{3.9b}$$

so that the first time-derivative disappears from the equation for v if

$$\alpha = \frac{B}{2\epsilon} . \tag{3.9c}$$

Similarly, the x-derivative may be eliminated.

Space-like subcharacteristics. The subcharacteristics of (3.8) are the lines

$$Ct - Bx = \text{constant} ,$$

and, as discussed above, their direction is that of increasing t. Consider now the case when the initial data vanish outside a finite interval of the x-axis. The reduced equation would then have a solution which is of order unity outside the region of influence of the initial data. Using principles stated above we find that the reduced equation cannot give an acceptable (outer) approximation to the exact solution.[†] An important comment on this apparent paradox was given by Cole (1968) who proved that an equation with space-like subcharacteristics is unstable in the sense that discontinuities in the derivatives of u across a characteristic grow exponentially along the characteristic. More specifically Cole proved the following:

Assume that at a point with characteristic coordinates (r_0, s_0) the function u is continuous across the characteristic $r = r_0$ but its derivative $\frac{\partial u}{\partial r}$ is discontinuous. Define the discontinuity as $\delta = \frac{\partial u}{\partial r}(r_0+, s_0) - \frac{\partial u}{\partial r}(r_0-, s_0)$. Then

1a) $B - C > 0$ implies exponential decay of δ along $r = r_0$;

1b) $B - C < 0$ implies exponential growth .

[†] Geel (1978, Chapter III) considers the problem

$$\epsilon\left(u_{tt} + (c_1 + c_2)u_{xt} + c_1 c_2 u_{xx}\right) + a u_t + b u_x = 0 ,$$
$$u(x,0) = x, \quad v(x,0) = 0, \quad a_1, \ b, \ c_1, \ c_2 = \text{constant} .$$

The exact solution is

$$u = x - \frac{b}{a}t + \epsilon\frac{b}{a^2}\left(1 - e^{-\frac{at}{\epsilon}}\right) .$$

For $a > 0$ the outer limit of the exact solution solves the reduced equation with the correct boundary condition on $u(x_1, 0)$, no matter what the values of c_1 and c_2 are (they need not even be constant) and no matter what the subcharacteristics are. We regard this example as clever but freakish since it does not shed any light on the general situation. Note that u_{xt} and u_{xx} are identically zero and especially that $U(x)$ is $\neq 0$ except at $x = 0$. Our criterion for an acceptable method is that it must work when the initial data vanish outside a finite region.

Similarly, for the case of a discontinuity in $\frac{\partial u}{\partial s}$ across $s = s_0$ he proved

2a) $B + C > 0$ implies exponential decay of discontinuity :

2b) $B - C < 0$ implies exponential growth .

Combining these results Cole found that necessary and sufficient conditions for decay of a discontinuity in a derivative along any characteristic are

$$B > 0 \text{ and } \frac{B}{|C|} > 1 . \qquad (3.11)$$

The details of the proof are left to the reader (Exercise 3.1). We repeat that "along" a characteristic means in the direction of increasing r (or s), and that, furthermore, the existence of discontinuities in solutions of hyperbolic equations is not only legitimate but that any acceptable theory must be able to deal with such discontinuities.

Whitham (1974, Chapter II) studies the problem of exponential growth or decay of the solution itself with the aid of the Fourier and Laplace transforms.

Obviously certain limiting cases are left out in Cole's analysis; for instance, one characteristic direction might coincide with the subcharacteristic direction or we may have $B = C = 0$. We shall not discuss such limiting cases. (See, however, Exercise 3.2.)

Time-like subcharacteristics. To get a first idea of what the solutions of (3.8) look like, let us assume that the initial data U and V are constant. The exact solution is then independent of x, in fact, (3.8) becomes an ordinary differential equation for a mass-spring system. We have assumed $\epsilon > 0$ (small positive mass). The requirement $B > 0$ of (3.11) means positive damping. Finally, $A < 0$ (negative spring constant) implies instability. Thus we assume $A \geq 0$, $C = 0$ (for convenience) and $B = 1$. ($B > 0$ is necessary for positive damping and we may then adjust this constant to be unity; the case $B = 0$ is discussed in Exercise 3.2.)

We simplify further by assuming $A = 0$. The exact solution for constant initial data is then

$$u = U + \epsilon V \left(1 - e^{-\frac{t}{\epsilon}}\right). \tag{3.12}$$

This behavior is well-known from the study of initial-value problems for ordinary differential equations: the leading term of the outer solution satisfies only the first initial condition. To satisfy the second (on u_t at $t = 0$) we need a layer of thickness ϵ and also of amplitude ϵ. This adds a term ϵV to the outer solution.

We now let U and V vary with x. We find the beginning of the outer expansion to be

$$u = U(x) + \epsilon\left[tU''(x) + V(x)\right] + O(\epsilon). \tag{3.13}$$

This is the generalization of our special case (U and V constant), derived by Cole, who also gives the inner expansion (see Exercise 3.3). Since the x-axis is not a subcharacteristic we expect the transition layer to be of thickness ϵ. This turns out to be correct and agrees of course with (3.12).

de Jager and also Geel prove the correctness of (3.13) and of similar higher-order approximations. We have, however, two fundamental objections to the term $tU''(x)$. First of all, it keeps growing with t. Note also that de Jager and Geel restrict their proofs to compact domain which means, in particular, that t is bounded. However, one is often interested in the behavior for large values of t, for instance $t \sim \frac{1}{\epsilon}$. There is some discussion of this problem in Chapter 10 of Whitham's book and we shall return to it in the discussion of the telegraph equation below. Our second objection to the term $tU''(x)$ is that there may be discontinuities in $U(x)$ or its derivatives. These should be propagated along characteristics, not along subcharacteristics as indicated by (3.13). Assuming $U(x)$ to be, say, infinitely often differentiable, as is done in some of the proofs by de Jager and by Geel, amounts to ignoring a very fundamental aspect of the hyperbolic equation. To emphasize these problems further we shall now consider the telegraph equation,

$$abu + (a + b)u_t + u_{tt} - u_{xx} = 0 \ . \tag{3.14}$$

(See Courant-Hilbert Vol. II, pp. 192–193.) As remarked earlier in this section we may eliminate u_t. We obtain

$$-\frac{(a - b)^2}{4}v + v_{tt} - v_{xx} = 0 \ , \tag{3.15a}$$

with

$$v = e^{\frac{(a+b)t}{2}} u \ . \tag{3.15b}$$

In particular we get the general solution of

$$u = e^{-at}\left[f(x - t) + g(x + t)\right] \ , \tag{3.16a}$$

if

$$a = b \ , \tag{3.16b}$$

(Courant-Hilbert and Garabedian also discuss the solution of the general case, but the special case (3.16b) is sufficient for our purposes.) The discussion of this equation is relegated to Exercises 3.4 and 3.5.

Exercises for §3

3.1. Using standard theory of hyperbolic equations, prove (3.11).

3.2. Consider the case $A = 1$, $B = 0$, $C = 0$. Decide whether it is possible to introduce an equation containing v_t by a transformation $u = e^{-at}v$. Discuss this question.

3.3. Derive (3.13) and the matching part of the inner expansion.

3.4. Consider the telegraph equation (3.14) with $a = b$ and choose the constants so that the wave operator, but not all terms, are multiplied by ϵ. Try a layer-technique solution and compare it with the exact solution (3.16).

3.5. Consider a radiation problem with initial conditions $U = V = 0$ and the radiation condition $u(Q,t) = W(t)$ for $t > 0$ and $= 0$ for $t = 0$. Give the leading term of an asymptotic solution for the equations

$$u_x + 2u_t + \epsilon(u_{tt} - u_{xx}) = 0 , \qquad\qquad (a)$$

$$-u_x + 2u_t + \epsilon(u_{tt} - u_{xx}) = 0 , \qquad\qquad (b)$$

$$u_t + \epsilon(u_{tt} - u_{xx}) = 0 . \qquad\qquad (c)$$

3.6. Give an approximate solution of the initial value problem

$$u_t + \epsilon(u_{tt} - u_{xx}) = 0 ,$$

$$U(x) = 1 \text{ for } x < 0 , \quad U(x) = 0 \text{ for } x > 0 ,$$

$$V(x) = 0 .$$

Show that ϵ is an artificial parameter.

Discuss qualitatively this problem with the change

$$U(x) = 1 \text{ for } |x| < 1 ,$$

$$U(x) = 0 \text{ for } |x| > 1 .$$

3.4. Evolution Equations. Shocklike Phenomena

The discussion of parabolic equations will be centered around one phenomenon, namely that of shocks. We consider equations in a time variable t, and one space variable, x. The reduced equation will be of first order and its solution may exhibit discontinuities, which then for the solution of the full equation will have to be replaced by interior layers of rapid but continuous change. Various possibilities arise depending on whether the reduced equation is linear or not and on what the order of the full equation is.

A linear second-order equation. We consider the equation and initial condition

$$u_t + u_x = \epsilon u_{xx} , \quad 0 < \epsilon \ll 1 , \tag{4.1a}$$

$$u(x,0) = U(x) . \tag{4.1b}$$

This problem is to be solved for $t \geq 0$, $-\infty < x < \infty$.

The solutions to the reduced equation are waves moving to the right (in the direction of increasing x) with unit speed and without changes in amplitude due to diffusion, attenuation, etc. A jump in the initial conditions will then propagate as a discontinuity. Consider a very simple case

$$U_0 = -1 \text{ for } x < 0, \quad U_0 = 1 \text{ for } x > 0 . \tag{4.2}$$

The first-order outer solution f_0 is then (we always assume $t \geq 0$)

$$f_0 = -1 \text{ for } x < t , \quad f_0 = 1 \text{ for } x > t . \tag{4.3}$$

By a Galilean transformation $y = x - t$ we can make the discontinuity occur at the stationary point $y = 0$. We introduce an inner coordinate $\tilde{y}$ and find for the inner solution g_0

$$\frac{\partial g_0}{\partial t} = \frac{\partial^2 g_0}{\partial \tilde{y}^2} , \quad \sqrt{\epsilon}\tilde{y} = x - t , \tag{4.4a}$$

with the matching conditions

$$g_0 = -1 \text{ at } \tilde{y} = -\infty , \quad g_0 = 1 \text{ at } \tilde{y} = \infty . \tag{4.4b}$$

The solution, which includes the outer solution, is then

$$g_0 = \text{erf } \frac{\tilde{y}}{\sqrt{4t}} = \text{erf } \frac{x - t}{(4\epsilon t)^{1/2}} . \tag{4.5}$$

This happens to be an exact solution of the equation (4.1). Here, however, the main interest is in the asymptotic method of finding it. The structure of the solution will be discussed in connection with the solution of the next problem.

A quasilinear second-order equation. We shall replace (4.1a) by the simple-looking quasilinear equation

$$u_t + u u_x = \epsilon u_{xx} , \tag{4.6}$$

which was introduced by Burgers (1948) as a model equation for turbulence and which is usually called the Burgers equation. The initial condition (4.1b) is retained. Cole (see Lagerstrom, Cole and Trilling, 1949, p. 151 ff) introduced (4.6) independently as an equation approximately valid for an actual physical phenomenon, namely the time-dependent development of a weak transonic shock wave. He also showed how the equation may be related to the heat equation by a linearizing transformation; the results were further elaborated in Cole (1951). Hopf (1950) found the same linearizing transformation independently; it is now known as the Cole-Hopf transformation. The method is similar to that used to transform a first-order Riccati equation to a second-order linear equation. If u satisfies (4.6) then v, as defined by

$$u = -2\epsilon \frac{v_x}{v} , \tag{4.7a}$$

satisfies

$$v_t = \epsilon v_{xx} . \tag{4.7b}$$

(or a linear equation which is easily transformed into (4.7b) by standard methods). We emphasize that the Cole-Hopf transformation of (4.6) is exact; it is not a linearization in the sense that one neglects small-amplitude nonlinear terms to obtain an approximating linear equation. Cole's application of (4.6) to weakly nonlinear sound waves was continued by Lighthill (1956); it is discussed in a much broader context by Whitham (1974).[†]

Before discussing specific critical-value problems we shall introduce a convenient intuitive interpretation of (4.6). It will be similar to the original physical problem discussed by Cole; however, we assume that the proper dimensional analysis has been done and that units have been chosen such that various constants are equal to unity. Equation (4.6) will be considered as an equation for one-dimensional sound waves. The variable u will be considered both as velocity and momentum. Equation (4.6) is then an equation of momentum balance. The change of momentum is due to transport with the transporting velocity u and to diffusion with diffusivity (viscosity) ϵ (*cf.* Sections 2 and 5).

We consider the transformation to a moving coordinate system

$$t \rightarrow s = t , \qquad (4.8a)$$

$$x \rightarrow y = x - X(t) . \qquad (4.8b)$$

Since u is a velocity it should transform as

$$u \rightarrow v = u - V(t) , \quad V(t) = X'(t) . \qquad (4.8c)$$

[†] Actually (4.6) and the transformation (4.7) were known earlier (see, *e.g.*, Forsyth, Vol. 6, Ch. 21) but had disappeared from the mathematical literature, along with many other useful concrete results. Recently nonlinear evolution equations, such as Burgers equation, the Korteweg-de Vries equation, *etc.*, which have special transformation properties, exact solutions and other unusual properties, have been studied intensively and with great success with the aid of a wide variety of mathematical techniques. The ideas and results of this at present very active field of mathematics are beyond the scope of this book.

Then (4.6) becomes (we again write "t" instead of "s"),

$$v_t + V'(t) + vv_y = \epsilon v_{yy} . \tag{4.9}$$

In the special case of $X(t) = V \cdot t$, $V = $ constant, the transformation (4.8) is a Galilean transformation and we see that (4.6) is Galilean invariant. This is to be expected from Cole's derivation of (4.6) as a physically valid equation of motion.

Governing the behavior of solutions of (4.6) are two opposing effects. The term ϵu_{xx} leads to a diffusion of momentum: if the nonlinear term uu_x is omitted or replaced by u_x, the momentum diffuses according to the simple heat equation as shown at the beginning of the present section. However, retaining this term leads to the Riemann steepening effect.[†] To see this effect we neglect the diffusion term; in other words we consider the reduced equation

$$u_t + uu_x = 0 . \tag{4.10}$$

The wave speed, that is the speed of propagation of u, is now amplitude dependent, in fact equal to u. Consider the initial value $U(x_0)$ at a point x_0. Then u is constant ($= U(x_0)$) along the line $x - x_0 = U(x_0)t$, which is a subcharacteristic. Assume, say, that $U(0) > 0$ and that $U'(x) < 0$ for $x > 0$. The greater x is, the lower the speed of the wave starting from x at $t = 0$. Since the large-amplitude waves travel faster than the small-amplitude the wave profile becomes steeper in the sense that u_x increases with t. If equation (4.10) is taken literally the solution may be triple-valued for sufficiently large t. The steepening and an actual triple-valuedness can be observed for the height of a water wave on a sloping

---

[†] This was discovered in Riemann's famous paper (1860) on sound waves of finite amplitude, a paper which laid the foundation for the modern study of nonlinear hyperbolic equations. A recent exhaustive study of waves in compressible fluids is given by Courant and Friedrichs (1948). A thorough treatment of various types of nonlinear waves is given in Whitham's book (1974) which in particular has a chapter devoted to Burgers equation and, incidentally, also discusses early research by Stokes on nonlinear waves.

beach. In the present case, however, u must be one-valued although the steepening
is real. Even if the diffusivity ϵ is very small the diffusive effect must eventually
offset the steepening effect as $\frac{\partial u}{\partial x}$ becomes very large. Riemann assumed in a
similar (although more complicated) problem that an actual discontinuity, called a
shock, occurs for $\epsilon = 0$ and determined the speed and magnitude of the shock from
conservation laws. In other terminology such a discontinuous solution is an outer
solution. To find it we make a transformation (4.8) such that the shock occurs at
$y = 0$. The function $X(t)$, and hence the shock speed $V(t)$ are as yet unknown.
Integrating (4.9) from $y_1 < 0$ to $y_2 > 0$ gives

$$\frac{\partial}{\partial t} 2 \int_{y_1}^{y_2} v \, dy + 2 \int_{y_1}^{y_2} V'(t) dy = v^2(y_1) - v^2(y_2) + \epsilon(v_y(y_2) - v_y(y_1)) \ .$$

Letting $y_1 \uparrow 0$, $y_2 \downarrow 0$ and denoting the limiting values of $v(y)$ by v_1 and v_2
respectively we find, as ϵ also tends to zero, that $v_1^2 = v_2^2$. From the qualitative description we deduce that the velocity must decrease across the shock in the
direction of increasing y . Thus we get the jump condition

$$v_2 = -v_1 \ , \tag{4.11a}$$

or

$$u_2 - V(t) = -(u_1 - V(t)) \ . \tag{4.11b}$$

This determines the shock velocity

$$V(t) = \frac{u_1 + u_2}{2} \ . \tag{4.12}$$

We interpret this formula as follows: Let $U(x_1) = u_1$, $U(x_2) = u_2$ where
$x_1 < x_2$ and $u_1 > u_2$. The subcharacteristics coming from $x = x_1$ and
$x = x_2$ meet at a certain point in the (x,t)-plane at which we get a double-valued
solution. This point gives us $X(t)$.[†] We assume that at the point $(X(t),t)$ the

---

[†] For a more detailed description of the formation of shocks see for instance Courant-
Friedrichs (1948, p. 107 ff.).

velocity decreases discontinuously from u_1 to u_2. The shock starts at the lowest value of t for which two subcharacteristics meet, after which its position is a function $X(t)$ whose time-derivative is the shock velocity $V(t)$.

We have thus found the outer solution. The reasoning implies that a shock is formed except when $U'(x)$ is everywhere ≥ 0.

For ϵ small we expect the shock discontinuity to be replaced by a layer of continuous but rapid change. In the linear problem discussed earlier discontinuities for $t > 0$ were due to discontinuities in the initial conditions and occurred along subcharacteristics. In the present problem discontinuities start where subcharacteristics with different slopes meet and the shock does not have a subcharacteristic direction. In analogy with the problems treated in Section 2 we might expect the scaling parameter for the transition layer to be ϵ rather than $\epsilon^{1/2}$ and that the governing equation for the layer solution g_0 is an ordinary differential equation. In fact, multiplying (4.9) by ϵ we see that if we introduce

$$\tilde{y} = y\epsilon^{-1} = (x - X(t))\epsilon^{-1} , \qquad (4.13)$$

into (4.9) then $g_0(\tilde{y})$, the leading term of the inner expansion of $v = u - V(t)$ obeys the equation

$$g_0 \frac{dg_0}{d\tilde{y}} = \frac{d^2 g_0}{d\tilde{y}^2} , \qquad (4.14a)$$

with the matching conditions

$$g_0(-\infty) = v_1 , \quad g_0(+\infty) = -v_1 , \qquad (4.14b,c)$$

where v_1 is the value of the outer solution for v just ahead of the shock. Thus (note the similarity with the interior layers studied in Chapter II, Section 4)

$$g_0 = -v_1 \tanh \frac{v_1}{2} \tilde{y} . \qquad (4.15)$$

Validity of asymptotic solution higher order equations. The validity of the solution obtained above may be verified by use of the Cole-Hopf transformation. Since this is

an exact technique which may fail when (4.6) is modified slightly we shall verify that
the solution has at least an internal consistency in the sense that a layer-solution
can be found which matches the outer solution. This means that (4.14) can be
solved. Furthermore, to verify this we shall not use the fact that (4.14a) has an
exact solution or even that it may be studied by phase-plane methods. We shall
use only a crude plausibility argument: As $\tilde{y}$ tends to $-\infty$, g_0 tends to $v_1 > 0$. For
large negative $\tilde{y}$ we therefore approximate (4.14a) by the equation

$$v_1 \frac{dg_0}{d\tilde{y}} = \frac{d^2 g_0}{d\tilde{y}^2} . \tag{4.16}$$

This clearly has a solution which decays to v_1 exponentially as $\tilde{y} \to -\infty$. By a
similar argument we make it plausible that there is a solution which decays exponentially to $-v_1$ as $\tilde{y} \to +\infty$.

Consider instead the Korteweg-de Vries equation in the form obtained by replacing u_{xx} in (4.6) by u_{xxx}. We can reason as above and obtain (4.14) with
$\frac{d^2 g_0}{d\tilde{y}^2}$ replaced by $\frac{d^3 g_0}{d\tilde{y}^3}$. In this case our plausibility argument would fail. In fact,
an exact (and rather difficult) analysis by Lax shows that shocks do not occur in
this case.

Exercise for §4

Consider the solution of the equation (4.6) for $x > 0$, $t > 0$ subject to the
conditions

$$u(0,t) = 1 = \text{ const}, \ u(x,0) = 0 . \tag{a}$$

(This is called a "piston problem" in fluid dynamics.) Cole (1951) found, using
(4.7), that the exact solution is

$$u = \frac{\text{erfc}\left(\frac{x-t}{\sqrt{4\epsilon t}}\right)}{e^{1/2\epsilon \left(x - \frac{t}{2}\right)} \text{erf} \frac{x}{\sqrt{4\epsilon t}} + \text{erfc} \frac{x-t}{\sqrt{4\epsilon t}}} . \tag{b}$$

The following expression (see Lagerstrom, 1964, pp. 207, 208),

$$g_0 = 1/2 \left[1 - \tanh \frac{1}{4\epsilon} \left(x - \frac{t}{2} \right) - 2\epsilon \, ln \, 2 \right] , \qquad (c)$$

has the property $\lim_{\epsilon \downarrow 0} \frac{|u - g_0|}{\epsilon^n} = 0$ for all n uniformly, if an arbitrarily small but fixed neighborhood of the origin is excluded. Discuss, using the ideas of an artificial parameter introduced elsewhere in this chapter, why the layer technique does not give the phase shift $2\epsilon \, ln \, 2$.

3.5. Examples from Fluid Dynamics

Historical Remarks. Fluid dynamics has played a crucial role in the development of the layer-type perturbation techniques. While there are early (dating as far back as Laplace; see Chapter II Section 6) isolated instances of successful applications of this technique, none of them led directly to any further use of the technique or to its further development. For practical purposes the use of the technique started with Prandtl's introduction of the boundary-layer concept in his famous paper (1905) on fluids with very small viscosity. Of course, in the light of later research certain ideas of Prandtl's about boundary layers may be criticized. However, his pioneering paper introduced some essential ideas and techniques correctly; for instance, the technique of matching (to lowest order). Furthermore, the idea of boundary layers became an important part of fluid mechanics. Even so, it was regarded for a long time as a fixed, specialized technique for getting approximate answers to a very restricted class of problems.[†] It was not until around 1940 that mathematicians interested in fluid dynamics, notably K. O. Friedrichs and W. Wasow, embarked on successful attempts to look at boundary-layer theory from the general point of view

---

[†] Some contributions by S. Goldstein were, however, an important exception; see Lagerstrom (1975).

of asymptotic expansions. The very simple but still very instructive model example discussed in Chapter II, Section 1 was introduced by Friedrichs (1943) with the express purpose of justifying and explaining some ideas by Prandtl, especially his matching technique.[‡] Boundary-layer theory is part of the theory of flow at high Reynolds number (this term will be explained below). The perturbation theory of flow at low Reynolds numbers, initiated by Stokes (1850), led to some positive results but also to various unexplained paradoxes; it seemed to bear no resemblance to the theory of flow at high Reynolds numbers. However, by a profound rethinking of the ideas underlying the theory of Prandtl and its development by Friedrichs and others, Kaplun (1957) found the surprising result that flow at low Reynolds numbers can be discussed systematically (and without paradoxes) by using the method of matched asymptotic expansions.

Perturbation problems of fluid mechanics are not only of great historic but also of intrinsic interest. This will be illustrated by the discussion of some selected problems below. We have tried to avoid duplication with the extensive treatment of this subject given by Van Dyke (1964, 1975).

Dimensional Analysis of Viscous Incompressible Flow. In Chapter I Section 3 dimensional analysis was used for discussing a mass-spring system, especially for various limiting cases. While an intuitive physical terminology was used in many subsequent examples it was assumed, with the exception of Chapter II, Section 6, that the dimensional analysis had already been done. The problems in fluid dynamics to be discussed below will give us a welcome opportunity to show again how dimensional analysis may be important in discussing singular-perturbation problems.

[‡] It is interesting to note that Friedrichs (1953) was the first one to realize that Prandtl's ingenious theory of high-aspect ratio wings, of great practical importance in engineering, actually depends essentially on the technique of matched asymptotic expansions (see Van Dyke, 1975, p. 239 ff.).

We shall carry out the dimensional analysis for viscous incompressible stationary two-dimensional flow. The extension to three dimensions and to nonstationary
flow is obvious; the extension to the compressible case is considerably more involved; see Lagerstrom (1964, pp. 168-171, and also pp. 187-201). The dimensional
variables are

$$\underline{x}_d = (x_d, y_d) = \text{Cartesian coordinates}, \tag{5.1a}$$

$$\underline{u}_d = (u_d, v_d) = \text{Cartesian velocity components}, \tag{5.1b}$$

$$p_d = \text{pressure} . \tag{5.1c}$$

The dimensions are obvious, but it should be kept in mind that a two-dimensional
object is regarded as a cylinder parallel to a z_d-axis. A line segment in the (x_d, y_d)-
plane may thus be regarded as an area element if extended a unit distance in the
z_d-direction. Note that the dimension of pressure is then force per unit area.

In addition there are certain constant parameters[†] of the fluid (below "[ ]"
means dimension of):

$$\rho_d = \text{Density}, \ [\rho_d] = \text{mass/length}^3, \tag{5.2a}$$

$$\mu_d = \text{viscosity}, \ [\mu_d] = \text{momentum/length}^2, \tag{5.2b}$$

$$\nu_d = \mu_d \rho_d^{-1} = \text{kinematic viscosity}, [\nu_d] = \text{velocity} \cdot \text{length} . \tag{5.2c}$$

In addition, the boundary conditions will supply certain parameters,

$$U_\infty = \text{speed at infinity}, \tag{5.3a}$$

[†] Here temperature is not even mentioned; it is also considered constant. It seems to
contradict any conceivable thermodynamic equation of state to assume variable pressure
and constant temperature and density. However, as shown in Lagerstrom (1964, p. 187 ff.),
this assumption is actually valid as a lowest-order approximation if the local velocities are
small compared to the speed of sound and if the differences between temperature parameters occurring in the boundary conditions are small. Under those conditions the viscosity
may also be regarded as constant. In short, for the problems studied here thermodynamics
will play no role. There will be no independent equation for the energy balance. The role
of pressure is purely dynamic, not thermodynamic.

$$L = \text{characteristic body length} . \tag{5.3b}$$

(There may be no characteristic body length or there may be several.)

Under the assumptions made above the flow is governed by the Navier-Stokes equations

$$u_d \frac{\partial u_d}{\partial x_d} + v_d \frac{\partial u_d}{\partial y_d} + \frac{1}{\rho_d} \frac{\partial p_d}{\partial x_d} = \nu_d \left(\frac{\partial^2 u_d}{\partial x_d^2} + \frac{\partial^2 u_d}{\partial y_d^2} \right) , \tag{5.4a}$$

$$u_d \frac{\partial v_d}{\partial x_d} + v_d \frac{\partial v_d}{\partial y_d} + \frac{1}{\rho_d} \frac{\partial p_d}{\partial y_d} = \nu_d \left(\frac{\partial^2 v_d}{\partial x_d^2} + \frac{\partial^2 v_d}{\partial y_d^2} \right) , \tag{5.4b}$$

$$\frac{\partial u_d}{\partial x_d} + \frac{\partial v_d}{\partial y_d} = 0 . \tag{5.4c}$$

The first two equations describe the balance between transport of momentum and changes in momentum due to fluid stresses. The last equation, commonly called the continuity equation, describes the conservation of mass. It is satisfied identically if the velocities are derived from a streamfunction ψ_d ,

$$u_d = \frac{\partial \psi_d}{\partial y_d} , \quad v_d = -\frac{\partial \psi_d}{\partial x_d} . \tag{5.5a, b}$$

The velocities (u_d, v_d) are tangent to the lines $\psi_d = $ constant, called streamlines. The vorticity ω_d is defined by

$$\omega_d = \frac{\partial v_d}{\partial x_d} - \frac{\partial u_d}{\partial y_d} = -\frac{\partial^2 \psi_d}{\partial x_d^2} - \frac{\partial^2 \psi_d}{\partial y_d^2} . \tag{5.6}$$

By taking the curl of the momentum equations one finds

$$u_d \frac{\partial \omega_d}{\partial x_d} + v_d \frac{\partial \omega_d}{\partial y_d} = \nu_d \left(\frac{\partial^2 \omega_d}{\partial x_d^2} + \frac{\partial^2 \omega_d}{\partial y_d^2} \right) . \tag{5.7}$$

Note the similarity of this equation with the simple equations studied in Section 2: The vorticity is transported along the streamlines while at the same time it diffuses like heat, the kinematic viscosity playing the role of heat conductivity. If the vorticity and the transporting velocity were independent of each other (as they are for

the Oseen equation, to be discussed below) the analogy would be very close. However, in the present case the streamfunction, and hence the velocity, depends on
vorticity according to (5.6) which is Poisson's equation if the vorticity is regarded
as known. This nonlinearity makes problems using (5.4) much more complicated
than the simple ones discussed in Section 2.

In the examples to be discussed below we shall consider mainly stationary flow
past a finite or semi-infinite object ("body"). The so-called no-slip condition will
be imposed:

$$\text{Velocity is zero at the surface of the body.} \tag{5.8}$$

In a typical problem we also have boundary conditions at infinity. The essential
one is

$$\text{velocity at infinity} = (U_\infty, 0) \ . \tag{5.9a}$$

The constant U_∞ is called the free stream velocity. In addition, we have

$$\text{pressure at infinity} = 0 \ . \tag{5.9b}$$

This is a normalization rather than a boundary condition. It is obvious from $(5.4a, b)$
that the absolute level of the pressure is irrelevant, only pressure differences matter.
We assume of course that pressure at infinity is constant and are then free to choose
any value of this constant. Equation $(5.9b)$ represents the simplest choice. Finally,
the values of the constant parameters described in (5.2) must be given; these values
could be regarded as boundary conditions at infinity.

It may be easily verified that from the various dimensional parameters mentioned above one can normally[†] form only one nondimensional parameter,

$$\text{Reynolds number} = Re = \frac{U_\infty L}{\nu_d} \ . \tag{5.10}$$

[†] If the body has two characteristic lengths their ratio is another nondimensional parameter, called aspect ratio, or slenderness, *etc.*, depending on the specific problem. If the
body has no characteristic length there is no *overall* Reynolds number. The coordinates
defined by $(5.13a)$ are sometimes referred to as *local* Reynolds numbers.

It may be viewed as the geometric length L divided by the viscous length $\nu_d U_\infty^{-1}$.

The obvious way to form a nondimensional velocity is to divide by U_∞ ,

$$\underline{u} = (u, v) = \underline{u}_d U_\infty^{-1} . \tag{5.11}$$

A nondimensional pressure may be defined by

$$p = \frac{p_d}{\rho_d U_\infty^2} . \tag{5.12a}$$

We shall also have occasion to use

$$\tilde{p} = Re \; p = \frac{p_d L}{\mu_d U_\infty} . \tag{5.12b}$$

If the problem has a viscous length and one geometric length, there are two obvious ways of introducing nondimensional coordinates

$$\underline{x} = (x, y) = \underline{x}_d U_\infty \nu_\infty^{-1} , \tag{5.13a}$$

and

$$\underline{\tilde{x}} = (\tilde{x}, \tilde{y}) = \underline{x}_d L^{-1} = \underline{x} Re^{-1} . \tag{5.13b}$$

Equations (5.4) now may be written,

$$u \frac{\partial u}{\partial x} + v \frac{\partial u}{\partial y} + \frac{\partial p}{\partial x} = \frac{\partial^2 u}{\partial x^2} + \frac{\partial^2 u}{\partial y^2} , \tag{5.14a}$$

$$u \frac{\partial v}{\partial y} + v \frac{\partial v}{\partial y} + \frac{\partial p}{\partial y} = \frac{\partial^2 v}{\partial x^2} + \frac{\partial^2 v}{\partial y^2} , \tag{5.14b}$$

$$\frac{\partial u}{\partial x} + \frac{\partial v}{\partial x} = 0 , \tag{5.14c}$$

or

$$Re \left(u \frac{\partial u}{\partial \tilde{x}} + v \frac{\partial u}{\partial \tilde{y}} + \frac{\partial p}{\partial \tilde{x}} \right) = \frac{\partial^2 u}{\partial \tilde{x}^2} + \frac{\partial^2 u}{\partial \tilde{y}^2} , \tag{5.15a}$$

$$Re \left(u \frac{\partial v}{\partial \tilde{x}} + v \frac{\partial v}{\partial \tilde{y}} + \frac{\partial p}{\partial \tilde{y}} \right) = \frac{\partial^2 v}{\partial \tilde{x}^2} + \frac{\partial^2 v}{\partial \tilde{y}^2} , \tag{5.15b}$$

$$\frac{\partial u}{\partial \tilde{x}} + \frac{\partial v}{\partial \tilde{y}} = 0 . \tag{5.15c}$$

Note that in $(5.15a, b)$ Re does not multiply the pressure gradient explicitly if one introduces $\tilde{p}$ from $(5.12b)$.

Flow at Low Reynolds Numbers.[†] One reason for starting with this problem is that it gives us an opportunity to introduce the Oseen equations and discuss their significance as approximating equations as opposed to that of model equations. The problem will also illustrate how dimensional arguments combined with dimensional analysis can be used effectively in heuristic arguments.

The definition of Reynolds number (5.10) and the boundary conditions described earlier show that one may let Reynolds number tend to zero by letting one of the dimensional parameters tend to zero or infinity while keeping the other parameters fixed. In the older literature dimensional terminology often predominated. Thus low Reynolds number flow was often described as slow flow and it was argued that since velocity was small one could linearize in the velocity. Since the first two terms of $(5.4a)$ and $(5.4b)$, called transport terms or inertia terms, are quadratic in the velocity they would then be omitted. The resulting linearized equations, now called the Stokes equations, were introduced by Stokes (1851). However, the reason given above for linearization is not valid since at infinity the nondimensional velocity is $\underline{u}_\infty = \underline{i}$. Thus the nondimensional velocity is not uniformly small. Of course, it is small near the body because of the no-slip condition but this condition is independent of Reynolds number. One might make a more convincing case for the Stokes equations if one regards the flow as very viscous. It may then be

---

[†] The discussion below will put great emphasis on heuristic ideas; for analytical details and comparison with experiments and numerical computations the reader is referred to Van Dyke (1975). The reader is also urged to compare the discussion here with the discussion of the model problems of Chapter II Section 5. The heuristic arguments given for the model problems resemble the ones given here except that now we deal with a real physical problem and make extensive use of dimensional analysis. Because of their comparative simplicity we gave a detailed analytical discussion of the model problems using techniques which in principle could be applied to the present problem although we would get fewer results in closed form. See also Example B3 of Section 2.

argued that the inertia terms may be neglected compared to the viscous stresses. We then again obtain the Stokes equations; note that the pressure gradient cannot be neglected since we would then have three equations for two unknowns. Let us now consider the limit of viscosity tending to infinity at a fixed point x_d , keeping free-stream velocity and the body length fixed.[†] Using the analogy between heat conduction and viscosity we expect for two-dimensional flow that the influence of the no-slip condition increases so that the velocity at the point x_d tends to zero. In nondimensional terms the velocity at a point with $\tilde{x}$ fixed tends to zero with Reynolds number. However, this limit is evidently nonuniform since the velocity at infinity remains $\underline{u}_\infty$. We shall call this limit the inner limit. Finally, let the body length L tend to zero. To fix the ideas, assume that we consider flow past a finite object which shrinks to the origin as L tends to zero. The influence of the body at a point with fixed coordinate x_d will then diminish and eventually disappear. Thus the velocity at x_d will tend to the free-stream velocity. In nondimensional terms the coordinate x is fixed as Re tends to zero. This will be called the outer limit. Thus the outer limit of $\underline{u}$ is $\underline{u}_\infty$, but this limit is also nonuniform since $\underline{u}$ remains zero at the surface of the body. If we keep $\tilde{p} = Re \cdot p$ fixed in (5.15) we see that the inner limit of the Navier-Stokes equations gives the Stokes equations; the body length, expressed in $\tilde{x}$-coordinates is unchanged. The outer limit gives back the full Navier-Stokes equation but the body length expressed in x-coordinates is Re and hence tends to zero.

So far we have found two important limits. In each limit $\underline{u}$ tends to a constant (for the two-dimensional problem) and thus solves the corresponding equation (assuming the limiting pressure to be zero). However, there is no possibility of matching the constants and we have found nothing which resembles an asymptotic solution.

[†] Thus the Reynolds number tends to zero. We again use a reasoning in nondimensional quantities while retaining the intuitive advantage of thinking in terms of dimensional ones.

Before proceeding further we first give a few historical facts. The linearized Stokes equations described above were introduced by Stokes in 1850. He found that for the two-dimensional problem the equations have no solutions satisfying all boundary conditions; any nontrivial solution satisfying the boundary conditions at the body must grow at least logarithmically at infinity; this is called the Stokes paradox. However, Stokes found a solution for flow past a sphere in three dimensions. Whitehead (1889) investigated this solution by iterating on the Stokes approximation, and found again a logarithmic singularity at infinity; this became known as the Whitehead paradox. Oseen (1910) found the source of the difficulty: The Stokes equations are not uniformly valid at large distances. In particular he showed that if one analyzes the Stokes solution for the sphere it turns out that at large distances the inertia terms, neglected in the Stokes equations, are larger than the viscous terms. Oseen's analysis is now considered correct. He also proposed a way of solving the difficulties. Since the inertial terms are important at large distances and since the velocity there tends to $\underline{u}_\infty = i$ he proposed not to omit the inertia terms completely but to replace the operator $u\frac{\partial}{\partial x} + v\frac{\partial}{\partial y}$ by $\frac{\partial}{\partial x}$. The resulting equations are now known as the Oseen equations. As is easily seen the Oseen equations are obtained by linearizing around the velocity at infinity. Note that, in the terminology used in Section 2, the role of velocity as the (nondimensional) momentum density and as transporting velocity have been separated. The velocity $\underline{u}$ is still the nondimensional momentum density but the transporting velocity is $\underline{u}_\infty$ in the Oseen equations whereas it is $\underline{u}$ itself in the Navier-Stokes equations. The Oseen equations met with success.[†] Oseen himself solved his equations for the sphere and found a solution (see below) which seemed an improvement on the Stokes solution. Lamb (1911) found a solution for flow past a circular cylinder. Still there were

---

[†] Using a faulty reasoning Oseen arrived at the erroneous conclusion that his equations were valid for large Reynolds numbers. This is now only of historical interest.

some questions left open (although seldom asked). It is plausible that the Oseen
equations are valid at some distance from the body, but why are they uniformly
valid? And how do they fit into a systematic asymptotic scheme? A satisfactory
answer to those questions was finally given by Kaplun approximately one hundred
years after the publication of Stokes' paper mentioned above.[†] Our discussion here
follows Kaplun's ideas (published and unpublished).

We have given reasons for regarding the outer and inner limits as important
limits. But is one more important than the other, in other words is there some sort
of hierarchy? It must be remembered that even in a singular perturbation problem
something is being perturbed. The unperturbed solution has a certain priority,
even if in a singular perturbation problem the change due to the perturbation may
be large in a small region. Oseen pointed out correctly that the Stokes equations
are not uniformly valid at large distances. This might suggest that there should
be a boundary-layer correction near infinity. However, this is a false lead. We
recall the various limit processes discussed in dimensional terms above. The Stokes
limit corresponds to infinite viscosity, the Oseen limit to no body present. One of
Kaplun's key ideas was that the unperturbed state is that of uniform flow. The
perturbation consists in introducing a very small body into the fluid. Thus the
boundary layer is not at infinity but in a small region near the body where the
relative change in velocity is large. While various heuristic arguments might suggest

---

[†] Kaplun's method for solving the low Reynolds number problem was first published
in condensed form in Lagerstrom and Cole (1955, section "Method of Kaplun," pp. 873-
876). Though the technique of solution was described there the underlying ideas were
not elaborated. The discussion in Kaplun (1957) and Kaplun and Lagerstrom (1957) was
still rather condensed, and Proudman and Pearson (1957) were concerned with applying
the technique (in particular they resolved the Whitehead paradox) rather than with the
underlying ideas. Since an up-to-date discussion of analytical results and comparison with
experiments and numerical computations is given by Van Dyke (1975) our discussion is
mainly concerned with the basic ideas. Many of these ideas are also discussed in detail for
the model problem of Chapter II Section 5.

this, the main justification is that this assumption leads to a successful theory. Thus we regard the first term of the outer expansion to be the constant velocity $\underline{u}_\infty$ (and pressure $= 0$). It satisfies the outer boundary conditions and the full equations (which are unchanged if the outer limit is applied). Note that the next term in the outer expansion should then satisfy equations obtained by linearizing the velocity around $\underline{u}_\infty$, that is the Oseen equations. If we now tentatively assume Lamb's solution to be correct we find that in the limit for which $x Re^{-a}$ $(0 \leq a \leq 1)$ is constant as Re tends to zero the (nondimensional) velocity at any given point tends to the constant velocity $(1 - a > 0)$. The corresponding resulting equations for $0 < a \leq 1$ are the Stokes equations and Kaplun argued that there should be a *solution* of the Stokes equations which contains all these limits. Forgetting about Lamb's solution (since we do not yet know whether the Oseen equations are uniformly valid) we observe that if one lets $x\eta^{-1}$ be constant as Re tends to zero one obtains the Stokes equations as long as $Re \precsim \eta(Re) \precsim 1$. Using the heuristic principle of Chapter I we then assume there is a solution of the Stokes equation valid in the same domain. This solution would have a domain of overlap with the outer solution $\underline{u} = $ constant $= \underline{u}_\infty$ because of the extension theorem. The details of the matching are best seen by discussing two and three dimensions separately.

Flow past a circular cylinder. To complement the qualitative reasoning above we shall present some quantitative results. We consider a circle centered at the origin, whose diameter[†] is taken as the characteristic length L. Using our previous definition of Reynolds number and of nondimensional variables we may write Lamb's solution of the Oseen equations for the velocity as

$$\underline{u} = \underline{i} + \epsilon \underline{g}(\underline{x}) + \underline{\ell}(\underline{x}), \quad \underline{i} = (1,0) = \underline{u}_\infty , \tag{5.16a}$$

[†] Sometimes the diameter is used to define the Reynolds number but the radius to define the inner coordinate $\tilde{x}_d$. With our conventional nondimensional radius of the circle is $\tilde{r} = 1/2$ or, equivalently, $r = Re/2$.

$$\epsilon = \left(\log \frac{8}{Re} + 1/2 - \gamma\right)^{-1}, \quad \gamma = \text{Euler's constant} = .5772, \qquad (5.16b)$$

$$\underline{g} = -2e^{x/2}K_0(r/2) + 2\nabla\left(e^{x/2}K_0(r/2) + \log r\right), \qquad (5.16c)$$

$$\underline{\ell} = -\tfrac{1}{8}\tilde{\nabla}\frac{\tilde{x}}{\tilde{r}^2}. \qquad (5.16d)$$

We have used both outer and inner coordinates; ∇ stands for $\left(\frac{\partial}{\partial x}, \frac{\partial}{\partial y}\right)$ and $\tilde{\nabla}$ for $\left(\frac{\partial}{\partial \tilde{x}}, \frac{\partial}{\partial \tilde{y}}\right)$. We shall leave to the reader to verify to what approximation the boundary conditions on the circle are satisfied and to analyze the formulae asymptotically (see Exercises 5.1–5.3). We observe that if the vector $\underline{\ell}$ is left out (5.16a) gives us (within a multiplicative constant representing the total force on the cylinder) the fundamental solution of the Oseen equation. The fundamental solutions in two or three dimensions were derived by Oseen and discussed in Oseen (1927); for a less formal discussion see Lagerstrom (1964).

Flow past a sphere. Using an analogy with heat distribution we argued that as viscosity tends to infinity the velocity at every fixed point tends to zero for flow past a two-dimensional object. The same result cannot be expected in three dimensions. Underlying various phenomena, even outside the problems discussed here, is the following simple heuristic principle about geometric attenuation: *The higher the dimension, the stronger the geometric attenuation.* This is easily explained with the aid of an example. Let r_k be the radial coordinate in k dimensions. Consider now the "sphere" in two dimensions, $r_2 = 1$, and the sphere $r_3 = 1$ in three dimensions and assume that some diffusive quantity such as temperature or, as in our case, zero momentum is prescribed on the surface of the sphere. To compare the two cases we regard $r_2 = 1$ as a circular cylinder in three dimensions. It is clear the surface of the circular cylinder has much greater effect on the surrounding medium than the surface of a sphere. Similar considerations apply of course for $k = 3, 4$ *etc.* It is because of geometric attenuation that the fundamental solutions of Laplace's equation in n dimensions decay faster with distance from the origin the larger n is.

For $n = 2$ it does not decay at all but increases logarithmically. Thus Dirichlet's exterior problem for Laplace's equation with the dependent variable u equal to zero on $r_n = 1$ and $u = 1$ at $r_n = \infty$ can be solved for $n = 3$ but not for $n = 2$. If we have a problem with radial symmetry, as in the model equation of Chapter II Section 5, we need not restrict n to be an integer. The attenuation phenomena can also be studied for Example B3 of Section 2. To return to the present case: In two dimensions the value of $\underline{u}$ decreases to zero as diffusivity increases to infinity and there is no solution of Stokes equation satisfying the conditions at $r_2 = 1$ and $r_2 = \infty$. This is no longer true in three dimensions. In that case the Stokes solution is

$$\underline{u} = \underline{i}\left(1 - \frac{3}{4\tilde{r}}\right) + \tilde{\nabla}\left(\frac{3}{8}\frac{\tilde{x}}{\tilde{r}} + \frac{1}{32}\frac{\partial}{\partial \tilde{r}}\frac{1}{\tilde{r}}\right) \ . \tag{5.17}$$

Thus the leading term of the inner expansion shows no Stokes paradox, but computing the next term leads to the Whitehead paradox. It seems plausible that if one writes the Navier-Stokes equations for n-dimensional flow and then constructs an inner expansion by iterating on the Stokes solution then a so-called paradox will occur, sooner or later. The larger n is the further the paradox is delayed, that is, the later it will occur in the expansion. This is easily verified analytically for the model equation of Chapter II Section 5 for the cases $n = 2, 3, 4$.

An approximate solution for Oseen flow past a sphere was found by Oseen (1911). Its inner limit gives the Stokes solution and a repeated inner limit gives a correction to the drag. For this solution and for higher-order terms the reader is referred to Van Dyke (1975). From the point of view of singular perturbations the same phenomena, with the analytical work greatly simplified, are shown by our model equation in Chapter II Section 5. Here we shall instead discuss certain basic qualitative aspects of the Oseen equations.

Validity of the Oseen equation. We first ask the question: When is the Oseen equation valid as an outer equation? The analysis given above shows that its validity is based on the assumption that the velocity at a given point $\underline{x}_d$ tends to its free-stream value as the characteristic body length L tends to zero. For this assumption to be valid the limiting body, obtained as L tends to zero, must not be able to influence the flow. We shall say that such a body has no arresting power, that is the no-slip condition imposed on it does not cause the flow to slow down at any point away from the body. Then uniform flow becomes the leading term of the outer expansion. The next term will be of some order $\eta(Re)$ which varies with the dimension but we always have $\eta \prec 1$. The equation for this term is then obtained by linearizing the velocity around the free-stream velocity; it is thus the Oseen equation. It may seem obvious that an object with $L = 0$ has no arresting power. However, L can be chosen in many ways. For a finite object L should measure its size and may be taken to be the diameter, say the maximum of the distance between all possible pairs of points on its boundary. As L tends to zero the body shrinks to a point; a two-dimensional point may be considered as a three-dimensional straight line. The arresting power is obviously zero. But consider now semi-infinite objects. As typical representatives we take the parabola (parabolic cylinder) and the paraboloid of revolution symmetric with respect to the freestream and located in the region $x \geq 0$ and passing through the origin (which is usually referred to as the "nose"). The diameter as defined above is infinite but we may choose L as the radius of curvature at the nose. The limiting configuration is then a semi-infinite line, the positive x-axis. In a different context we asserted above that an object of a given type (say a sphere) has larger surface area the smaller the dimension. The same principle can be applied again. A half-line in two dimensions is a half-plane when considered as a three-dimensional configuration (in fluid dynamics it is often called a semi-infinite flat plate). Its arresting power is obviously not zero.

Thus while the Oseen equation can be solved for flow past a parabola the solution has no value as an asymptotic approximation for low Reynolds number flow. The correct equation, taking the role of the Oseen equation, would be obtained by linearizing around the viscous solution past a semi-infinite flat plate. This solution, as will be discussed below, is known only incompletely. On the other hand, a half-line in three dimensions has no arresting power, thus one may use the Oseen equations in the study of flow past a paraboloid.[†]

The type of reasoning used above about "arresting power" may be applied to other physical situations. One may discuss the "heating power" *etc.* of an object with a characteristic length L in the limit of vanishing L. Whenever the influence of the body vanishes at a fixed point which is not on the limiting body one may use linearization about the value at infinity (of velocity, temperature, *etc.*) as asymptotic equations valid for the outer solution. The introduction of nondimensional variables, suitable for the outer expansion, has to be done as illustrated for our particular example. In particular, it follows that the Oseen equations for compressible flow may have asymptotic validity, in some sense, for flow at low Reynolds numbers when the limiting body has no arresting (or heating) power.[‡]

Assume now that the limiting object as L tends to zero has no arresting power. Then the heuristic reasoning given above makes a good case for the use of the Oseen equations as *outer* equations at low Reynolds numbers. However, the use of the Oseen equations in finding solutions for flow past a circular cylinder or a sphere

---

[†] It is interesting to note that, for reasons to be described later, the parabola and the paraboloid belong to the very small class of configurations which are amenable to a practical application of high Reynolds number technique. The paraboloid of revolution is the simplest object which can be studied systematically for flow at small and large Reynolds numbers. Various calculations have been carried out by Van Dyke; see Van Dyke (1975) and references given there.

[‡] This statement needs to be qualified; see Lagerstrom (1964). It may not be valid for supersonic flow. However it would apply to low Mach number flow past a body whose surface temperature differs from that of the fluid at infinity.

for which the error in the inner boundary condition is small when Re is small suggests that the equations should be uniformly valid. Another type of reasoning indicates that this is actually true: *For incompressible flow the Oseen equations contain (are richer than) the Stokes equations.* (For the discussion of this principle in a simple context, see the subsection "The Method of Kaplun" in Chapter II Section 5.) Indeed, applying the inner limit to the Oseen equations yields the Stokes equations. Correspondingly (see Exercises), applying the inner limit to the Oseen solutions yields the Stokes solutions. Thus the approach to the low Reynolds number problem advocated by Kaplun not only explains why the Oseen equations are outer equations but solves the puzzle of why they are valid near the body. It also showed clearly (see Kaplun-Lagerstrom, 1957) that uniform validity of the Oseen equations cannot be expected for compressible flow. *The Oseen equations are by nature linear*; an extension of the reasoning used here shows that one should linearize not only about the velocity at infinity but also in temperature, density, *etc.* The Stokes equations are obtained not by linearization but by neglecting the inertial terms relative to the stress terms or, more formally, by a limit process. Only when this limit happens to give linear equations is there a chance that the Stokes equations may be contained in the Oseen equations. However, if the former are not contained in the latter one cannot match the inner and outer solutions correctly. It was to illustrate this situation analytically that the term $b \left(\frac{du}{dx} \right)^2$ was added to the model equation (Chapter II Section 7). If one linearizes in u this term will disappear, say for $\beta = 1$. A solution of this equation cannot be matched with the outer solution (which satisfies a linear equation corresponding to the Oseen solution). The situation is actually even worse: matching is formally possible but gives the wrong result. However matching is possible if one obtains the inner equation by a correct limit process. The resulting equation (which corresponds to the Stokes equation) is then nonlinear and has a solution which can be matched correctly to the outer solution.

The Oseen equations may also be used for any Reynolds number as a starting point in discussing flow at large distances from a finite object. Boundary-layer techniques may be used for matching the inner viscous wake with the surrounding outer flow. In this way matched asymptotic expansions may be obtained. Matching with flow near the body is not possible[†] and the expansion will contain undetermined constants. Because of the existence of conservation laws of mass, momentum and angular momentum a few of these constants may be identified with certain integrated quantities on the surface of the object, namely total source strength, total force and total angular momentum. The total source strength is zero if the no-slip conditions are assumed but the other quantities are not known. An exhaustive discussion of this problem is given in Chang (1961).

Flow Past a Semi-infinite Flat Plate. Parameter and Coordinate-type Expansions. We consider two-dimensional viscous flow with the no-slip condition $u_d = 0$ and $v_d = 0$ imposed on the positive x-axis (the "flat plate"). As usual we assume the velocity at infinity to be $U_\infty i$ and nondimensionalize the velocity components by dividing by U_∞. However, the plate has no length parameter, hence there is no overall Reynolds number and there is only one way of introducing nondimensional coordinates, namely by (5.13). Then all parameters disappear from the resulting equations (5.15) as well as from the boundary condition. In particular, there is no parameter which tends to zero or to infinity; still this was the boundary value problem which Prandtl (1905) used to introduce the basic ideas of boundary-layer theory. The very simple Example B2 of Section 2 shows us the answer to this apparent paradox: In spite of the fact that viscosity is a very real physical parameter, for the present boundary value problem it must be considered an artificial param-

---

[†] The analysis uses an artificial parameter and the results are not uniformly valid. *Cf.* discussion of the semi-infinite plate below.

eter. However, we can construct matched inner and outer expansions with the aid

of this parameter, and the use of the dimensional parameter helps the heuristic rea-

soning. (Note the title of Prandtl's 1905 paper: "On fluid motion with very small

viscosity.") From Example B2 in Section 2 we learn one important fact: Prandtl's

solution is not uniformly valid for all x and y; more specifically it fails near the

origin. In the terminology of Lagerstrom and Cole (1955) the expansion obtained

is a coordinate expansion rather than a parameter expansion, being valid for large

values of $x = \frac{U x_d}{\nu_d}$. We note that for a given value of $x_d > 0$ the value of x

increases as ν_d decreases. In this sense the solution is really a solution for small

viscosity, as long as $x_d > 0$. In addition to the similarity found from dimensional

analysis the geometry of the problem gives an additional similarity which implies

that u must be a function of $y x^{-1/2}$ (cf. Example B2 of Section 2). Using this,

Prandtl reduced the boundary layer equations to an ordinary differential equation,

later called the Blasius equation, for which he gave an approximate solution. This

equation has since been treated extensively in the literature on fluid dynamics, to

which we refer the reader for a discussion of the solution. We shall discuss only

one point here. Prandtl's solution gives a viscous shearing force on the plate ("skin

friction") which is proportional to $1/\sqrt{x}$ and hence integrable for any finite inter-

val constant $0 \le x \le$ constant. Attempts to iterate on the Prandtl solution give

nonintegrable singularities at $x = 0$. This was once regarded as a paradox since

the total viscous shear on a finite front portion of the plate would be infinite. Like

numerous other paradoxes in fluid mechanics this one also resulted from a faulty

understanding of perturbation methods: The expansion to which Prandtl's solution

contributed the leading terms is a coordinate-type expansion and it cannot be uni-

formly valid unless we have the case of "exceptional luck" discussed in Example B2

of Section 2. The difference between coordinate-type expansions versus parameter-

type expansions are discussed in Lagerstrom-Cole (1955) and in great detail by

Chang (1961) who gives a systematic discussion (based on unpublished ideas by Kaplun) of how artificial parameters may be used in applying layer techniques to certain coordinate-type expansion problems.[†]

Higher terms of the coordinate-type expansion for the flat plate were calculated by Imai and Goldstein. For discussion and references, see Van Dyke (1975).

Oseen Equations as Model Equations. The Oseen equations can be used as model equations in cases for which they are not equations for asymptotic approximations. Let us, for instance, impose the same boundary values as in the previous example (semi-infinite flat plate) but use the Oseen equations instead of the Navier-Stokes equations. The boundary layer approximation in rectangular coordinates is easily found. By a tricky use of the Wiener-Hopf technique Lewis and Carrier (1949) found the exact solution using rectangular coordinates. A significant shortcut to the solution was found by Van Dyke (private communication, see Kaplun, 1951) who observed that if one finds the boundary-layer solution in parabolic coordinates one actually has the exact solution of the Oseen equations.[‡] This is thus a case of what we have called exceptional luck above. Since the flat plate is located along the x-axis it has a subcharacteristic direction. Thus, the boundary-layer is of thickness $\nu^{1/2}$. Furthermore, the geometric similarity in each case is the same. (See Exercise 5.4.) As a result, both the Prandtl boundary-layer solution and the exact solution of the Oseen equation give a shearing force along the flat which is proportional to $1/\sqrt{x}$ (where x is defined by Equation 5.13a), only the constant of proportionality differs. In this case, the Oseen solution seems quantitatively similar to the Navier-Stokes solution. That is, however, a coincidence which in particular depends on the

---

[†] Incidentally, elaborating a suggestion by Goldstein, Chang's paper gives a systematic resolution of another spurious paradox in fluid dynamics, the Filon paradox.

[‡] A similar phenomenon was encountered in Example B2 of Section 2, although Van Dyke's discovery was made independently of that example.

fact that the flat plate has subcharacteristic directions. In most cases a solution of the Oseen equation is even qualitatively different from the corresponding solution of the Navier-Stokes equations, except in those cases for which the Oseen equations are approximately valid for low Reynolds number. Take for example flow past a parabola. The Oseen solution is obtained by a variation of Van Dyke's method for solving the flat plate problem. However, as discussed above this solution has no value as an approximation for small Reynolds numbers. For large Reynolds numbers the boundary layer has thickness ν. The solution is then qualitatively different from that of the Navier-Stokes solution.[†]

Note on a fundamental difference between perturbation theory at high and low Reynolds numbers. The remarks below refer to a very basic aspect of perturbation theory. They are of general nature and might have been placed in the introductory Chapter I. However, we insert them at this point because in this book they are best illustrated by the preceding discussion of real physical problems.

The basic premise is that any physical theory is an approximation to a more accurate theory. For example, classical dynamics is an approximation to quantum mechanics and also to special relativity, Newtonian theory of gravitation is an approximation to Einstein's general theory of relativity, *etc.* Clearly the older theories should not be considered wrong since they generally give good agreement with observation, although at some stage or other more precise measurements or observations of new phenomena makes the introduction of more accurate theories desirable. Thus there must be a close connection between an older theory and a more accurate theory. This connection usually takes the following form: The older theory is the limit of the newer theory as a certain fundamental parameter tends to zero. For small values of this parameter the newer theory exhibits small correc-

---

[†] The Oseen solution can still be used to illustrate certain general ideas about inner, outer and composite expansions as was done in Lagerstrom (1964, p. 151 ff.).

tions to the older theory. These corrections are observed experimentally by very accurate measurements and theoretically, in principle at least, explainable by perturbation methods (regular or singular) in which the older theory gives the zeroth order approximation.[†] Examples of this abound but only one illustrative example will be cited. Newton's theory of gravitation has been shown to be a limiting case of Einstein's theory; accurate observations of the motion of the perihelion cannot be explained by Newton's theory alone, but by a perturbation of this theory using a small relativistic correction. This aspect of perturbation is clearly of fundamental importance for physics, and for any branch of science amenable to a mathematical formulation. The perturbation is singular if the effect of the correction is confined to a very small region, or if it is important over a very long time period as is the case for Mercury.

Against this type of perturbation problem we contrast another type which is not of basic importance for physics but rather has to do with practical methods of finding approximate formulas. Let us consider problems wholly within the framework of Newtonian gravitational theory. The two-body problem can be solved exactly. If the motion of a planet is slightly perturbed by the presence of other planets the result may be computed by perturbation methods. Clearly such problems do not involve a change in a basic physical theory, they only compare a relatively simple problem with a slightly more complicated problem within the same physical theory.

Returning now to the problems discussed earlier in this section we notice that there is a theory for zero viscosity. This theory is expressed by the Euler equations;[†] the Navier-Stokes equations express a more accurate theory. Thus the

---

[†] For large values of the parameter the new theory may give results which are even qualitatively new and which are beyond the scope of perturbation theory.

[†] As a theory it has many shortcomings, for instance solutions of boundary-value problems may be highly non-unique. This need not concern us here. We remark only that the mathematical investigations of the Euler equations are remarkably incomplete; as a

high Reynolds number problems belong to the first class of perturbation problems discussed above; solutions of the Euler equations should give zeroth order approximations to problems with small viscosity.

On the other hand, nobody ever proposed a theory of infinite viscosity. Thus in the low Reynolds number problems we work within the framework of one physical theory, namely that of viscous fluids. In the solvable problems of low Reynolds number flow we have a very simple solution of the Navier-Stokes equations, namely uniform flow, which is perturbed by the presence of the small body. In the small neighborhood of this body viscosity predominates and the Stokes equations are used. Thus the rather philosophical arguments given above confirm the results obtained earlier by more concrete reasoning: It is the free stream (a solution of the Navier-Stokes equation) which is being perturbed by a small change in the boundary conditions. Assuming the Stokes equations to be basic (which was done in the last century and led to paradoxes) amounts to assuming a theory of infinite viscosity (expressed by the Stokes equations) which is being corrected by a theory which allows viscosity to be finite.

After this discussion on the basic nature of perturbation problems we shall discuss a problem raised by various solutions given earlier.

The Role of Coordinate Systems in Boundary-layer Theory. The investigation of this problem was prompted by Van Dyke's observation about the Oseen solution for the semi-infinite flat plate. This and a few other concrete examples showed that different coordinate systems yield boundary-layer solutions which describe flow fields which are different but still described by almost similar expressions. Secondly, it was observed that some coordinate systems yielded better solutions than others. The word "better" will be defined below; it does not in general mean exact, as in

---

consequence the problem of finding the relevant solutions (see Section 1) is sometimes very difficult.

Van Dyke's case. This raised two questions to which remarkable simple and general answers were given by two theorems of Kaplun (1954). Kaplun's work refers specifically to boundary-layers in fluid flow. His first theorem can be trivially extended to much more general situations; the second (and more important) theorem has so far not been successfully liberated from its hydrodynamical context.

The first theorem is:

Kaplun's Correlation Theorem. *Let (x, y) and (s, t) be two coordinate systems completely general except for the requirement that the coordinates y and t vanish at the boundary where the layer occurs. Assume that near the boundary x and y can be developed in power series in t. Let $\eta(\epsilon)$ be the scaling parameter and define $\tilde{y} = y\eta^{-1}$, $\tilde{t} = t\eta^{-1}$. Assume that the boundary-layer solutions are obtainable from the exact solution $u(x, y; \epsilon)$ by inner limits. Then*

$$g_0^{(2)}(s, \tilde{t}) = g_0^{(1)}\left(x(s, o), \left(\frac{dy}{dt}\right)_{t=0} \tilde{t}\right). \qquad (5.18)$$

Here $g_0^{(1)}(x, \tilde{y})$ is the boundary-layer solution relative to the first coordinate system and $g_0^{(2)}(s, \tilde{t})$ the corresponding result for the (s, t)-system.

PROOF: Write $u(x, y; \epsilon) = g(x, \tilde{y}; \epsilon)$ and form the appropriate limits using the beginning of the series of developments of $\underline{x(s_1)}$ and $\underline{y(s_1 t)}$.   □

Comments. Obviously, the functions x and y need not be analytic in t, only very weak continuity or differentiability requirements are needed. The function g need only be continuous from the right in ϵ. Also, the theorem is valid, *mutatis mutandis*, for any number of independent variables. Kaplun's essential contribution was thus to realize that one needs only an almost trivial theorem about limits to obtain an important results which previously had gone unnoticed. Furthermore, his method eliminates any need for actually writing down the boundary-layer equations.[†] In

---

[†] These equations were actually written out in his paper (for different reasons) but not

fact, it shows that the result is in no way restricted to fluid dynamics. This is illustrated by Exercise 5.5 which also shows that the parameter used in forming the limit may be artificial.

As an easy corollary of the correlation theorem Kaplun gave necessary and sufficient conditions that two coordinate systems give the same boundary layer (Exercise 5.6).

As mentioned, Van Dyke obtained the exact solution for Oseen flow past a semi-infinite flat plate by solving the boundary-layer problem in parabolic coordinates. We have just seen that this solution could have been obtained by a direct substitution in the well-known solution in parabolic coordinates. However, the main point is that parabolic coordinates are obviously better than rectangular. On the other hand, if one applies the same procedure to the full Navier-Stokes equations one finds that parabolic coordinates do not yield the exact solution; however, the boundary-layer solution contains the first two terms of the outer expansion. Kaplun called such coordinates *optimal*. Kaplun's *second theorem* is the *Theorem on Optimal Coordinates* in which he gives remarkably simple necessary and sufficient conditions for coordinates to be optimal. For details we refer to Kaplun's original paper, reprinted in Kaplun (1967), and to Van Dyke (1964, 1975). We note that optimal coordinates are not determined *a priori*: One must first find the boundary-layer solution in some coordinate system, then find the next term in the outer solution (which is called flow due to displacement thickness and is a solution of Laplace's equation). Then optimal coordinates are easily found from Kaplun's second theorem and the solution in optimal coordinates can be written down using the correlation theorem. Thus Kaplun's two theorems enable us to write part of the composite expansion in a neat form.

---

used in his proof. In fluid dynamics $\psi_d \nu_d^{-1/2}$ (or in some nondimensional equivalent thereof) plays the role of u above.

Various extensions of Kaplun's theorems to higher order terms, to three-dimensional flows *etc.* have been given. We refer to Note 9 of Van Dyke (1975) and references given there. In Kaplun's original determination of optimal coordinates limit processes were used without direct reference to equations. However, his reasoning depended on the peculiar structure of boundary layers in fluid dynamics. None of the extensions of his second theorem have clarified what the essence of this structure is or made a nontrivial extension to problems outside fluid dynamics. Compare Exercise 5.5. In this problem Kaplun's correlation theorem is used to switch from a boundary-layer solution in rectangular coordinates. The latter solution happens to be the exact solution of the problem, but there is no general theorem to predict *a priori* why this is the case.

Exercises for §5

5.1. a) Take the inner (Stokes) limit of (5.16a), *i.e.*, replace x by $\tilde{x} = x\, Re^{-1}$ and let Re tend to zero keeping $\tilde{x}$ fixed. Show that this gives zero.

b) Then take the inner limit of $\underline{u}\epsilon^{-1}$ and obtain a vector $\underline{g}_0(\tilde{x})$.

c) Show that $\epsilon \underline{g}_0(\tilde{x})$ matches with $\underline{u}_0 = \underline{i}$ in a narrow domain (*cf.* Chapter II, Section 7).

5.2. Show that $\epsilon \underline{g}_0(\tilde{x})$ satisfies the Stokes equation and the inner boundary condition but not the condition at infinity (which, as shown in Exercise 5.1) is replaced by a matching condition.

5.3. Show that (5.16) gives a solution which is uniformly valid to a certain order. Evaluate the error at the surface of the cylinder.

5.4. Show that the Oseen equations (written in terms of a streamfunction) can be made into an ordinary differential equation for the semi-infinite flat plate. Do this first in rectangular and then in parabolic coordinates.

5.5. a) Show that the solution (2.33) is obtained directly from (2.30) with the aid
of Kaplun's correlation theorem.

b) Using rectangular coordinates find the solution of the boundary-layer equation for Oseen flow. Using the correlation theorem find the corresponding
solution in parabolic coordinates. By writing the Oseen solution in parabolic
coordinates show that the latter solution is an exact solution of the Navier-
Stokes equations.

5.6. Using Kaplun's correlation theorem show that the two coordinate systems (x, y)
and (s, t) lead to the same boundary-layer solution if and only if $x = f_1(s)$ and
$y = f_2(s)t$, where f_1 and f_2 are arbitrary functions. Interpret the preceding
equations geometrically in terms of shape of coordinate lines and their labelling.

REFERENCES

Ackerberg, R. C., and O'Malley R. E., Jr. (1970). Boundary layer problems exhibiting resonance. *Studies in Appl. Math.* **49**, 277–295. [70]

Bender, C. M., and Orszag, S. A. (1978). *Advanced Mathematical Methods for Scientists and Engineers.* McGraw-Hill, New York. [55,57,66]

Bush, W. B. (1971). On the Lagerstrom mathematical model for viscous flow at low Reynolds number. *SIAM J. Appl. Math.,* **20** (2), 279–287. [107,125]

Casten, R. G., Cohen, H., and Lagerstrom, P. A. (1975). Perturbation analysis of an approximation to the Hodgkin-Huxley theory. *Q. Appl. Math.,* **32**, 365–402. [162,163]

Chang, I. (1961). Navier-Stokes solutions at large distances from a finite body. *J. Math. Mech.,* **10**, 811–876. [4,193,232]

Chang, K. W., and Howes, F. A. (1984). *Nonlinear Singular Perturbation Phenomena: Theory and Application.* Springer-Verlag, New York. [86]

Cohen, D. S., Fokas, A., and Lagerstrom, P. A. (1978). Proof of some asymptotic results for a model equation for low Reynolds number flow. *SIAM J. Appl. Math.,* **35**, 187–207. [107,118,123,125]

Cole, J. D. (1951). On a quasilinear equation ocurring in aerodynamics. *Q. Appl. Math.,* **9**, 225–236. [86,160,202]

Concus, P. (1968). Static menisci in a vertical right circular cylinder. *J. Fluid Mech.,* **34**, part 3, 481–495. [134,495]

Corben, H. C., and Stehle, P. (1960). *Classical Mechanics.* John Wiley and Sons. [152]

Courant, R., and Friedrichs, K. O. (1948). *Supersonic Flow and Shock Waves.* Interscience, New York. [212,213]

Courant, R., and Hilbert, D. (1962). *Methods of Mathematical Physics, Vol. II.* Interscience, New York. [199,207]

de Jager, E. M. (1974) (Ed.). Spectral theory and asymptotics of differential equations. *Proc. of the Scheveningen Conference on Differential Equations,* September 3–7, 1973. [177]

de Jager, E. M. (1975). Singular perturbations of hyperbolic type. *Nieuw Arch. voor Wiskunde,* **23** (3), 145–171. [202]

Eckhaus, W. (1969). On the foundations of the method of matched asymptotic approximations. *J. de Mécanique,* **8**, 265–300. [9]

Eckhaus, W. (1972). Boundary layers in linear elliptic singular perturbation problems. *SIAM Review,* **14**, 43–88. [177]

Eckhaus, W. (1973). *Matched Asymptotic Expansions and Singular Perturbations.* North-Holland, Amsterdam. [71]

Erdélyi, A. (1956). *Asymptotic Expansions.* Dover New York. [55]

Erdélyi, A., Mahnus, W., Oberhettinger F., Tricomi, F. G. (1953). *Higher Transcendental Functions, Vol. II.* McGraw-Hill. [69]

Erdélyi, A., and M. Wyman (1963). The asymptotic evaluation of certain integrals. *Arch. Rat. Mech. Anal.,* **14**, 217–260. [8]

Finn, R. (1986). *Equilibrium Capillary Surfaces.* Series Grundlehren der mathematischen Wissenschaften 284, Springer-Verlag, New York-Berlin-Heidelberg-Tokyo. [135]

Fitz Hugh, R. (1961). Impulses and physiological states in theoretical models of nerve membrane. *Biophys. J.,* **1**, 445–466. [162]

Fokas, A. S. (1979). Group theoretical aspects of constants of motion and separable solutions in classical mechanics. *J. Math. Anal. Appl.,* **68**, No. 2, 347–370. [152]

Fokas, A. S., and Lagerstrom, P. A. (1980). Quadratic and cubic invariants in classical mechanics. *J. Math Anal. Appl.,* **74**, No. 2, 325–341. [152]

Forsythe, A. R. (1906). *Theory of Differential Equations.* Dover reprint. [211]

Fraenkel, L. E. (1969). On the method of matched asymptotic expansions, I–III. *Proc. Camb. Phil. Soc.,* **65**, 209–284. [80,81]

Freund, D. D. (1972). A note on Kaplun limits and double asymptotics. *Proc. Amer. Math. Soc.*, **35**, 464–470. [24]

Friedrichs, K. O. (1942). *Fluid Dynamics*. Chapter 4. Brown University. (Reprinted in Springer (1971).) [31,217]

Garabedian, P. R. (1964). *Parial Differential Equations*. John Wiley, New York. [199,207]

Geel, R. (1978). *Singular Perturbations of Hyperbolic Type*. Mathematisch Centrum: Amsterdam. [202,204,206]

Grasman, J. (1971). *On the Birth of Boundary Layers*. Mathematisch Centrum Amsterdam. [177,182]

Grasman, J. (1974). An elliptic singular perturbation problem with almost characteristic boundaries. *J. Math. Anal. Appl.*, **46**, No. 2, 438–446. [177]

Grasman, J., and Matkowsky, B. J. (1976). A variational approach to singularly perturbed boundary value problems for ordinary and partial differential equations with turning points. *SIAM J. Appl. Math.*, **32**, No. 3, 588–597. [68]

Hardy, G. H. (1924). *Orders of Infinity*, 2nd ed. Cambridge Tracts, No. 12. [5,6,9]

Hodgkin, A. F., and Huxley, A. F. (1952). A quantitative description of membrane current and its application to conduction and excitation in nerve. *J. Physiol.*, **117**, 500–544. London. [162]

Hopf, E. (1950). The partial differential equation $u_t + uu_x = \mu u_{xx}$. *Comm. Pure Appl. Math.*, **3**, 201–230. [369]

Hsiao, G. C. (1973). Singular perturbations for a nonlinear differential equation with a small parameter. *SIAM J. Math. Anal.*, **4** No. 2. [107]

Kaplun, S. (1954). The role of coordinate systems in boundary-layer theory. *Z. Angew. Math. Physics*, **5**, 111–135. [20,193]

Kaplun, S. (1957). Low Reynolds number flow past a circular cylinder. *J. Math. Mech.*, **6**, 595–603. (Reprinted in Kaplun (1967).) [111,124,217,225]

Kaplun, S. (1967). *Fluid Mechanics and Singular Perturbations*. P. A. Lagerstrom, L. N. Howard, and C. S. Liu (eds.), Academic Press: New York. [4,6,259]

Kaplun, S., and Lagerstrom, P. A. (1957). Asymptotic expansions of Navier-Stokes solutions for small Reynolds numbers. *J. Math. Mech.*, **6**, 585–593. (Reprinted in Kaplun (1967).) [29,41,107,111,225,231]

Kevorkian, J. (1962). The two variable expansion procedures for the approximate solution of certain nonlinear differential equations. Douglas report SM-42620, Douglas Aircraft Company, Inc. (Also presented at the 1962 Summer Institute in Dynamical Astronomy.) [173]

Kevorkian, J., and Cole, J. D. (1981). *Perturbation Methods in Applied Mathematics.* Springer-Verlag, New York. [19,55,62,64,68,86,98,107,157,160,161,202]

Kevorkian, J., and Lancaster, J. E. (1968). An asymptotic solution for a class of periodic orbits of the restricted three-body problem. *Astronomical J.*, **73**, No. 9, 791–806. [157]

Kreiss, H. O. (1981). Resonance for singular perturbation problems. *SIAM Appl. Math.*, **41**, No. 2, Philadelphia, PA. [70]

Krylov, N., and Bogoliubov, N. (1947). *Introduction to Nonlinear Mechanics.* Translated by S. Lefschetz, Princeton University Press. [172]

Kuzmak, G. E. (1959). Asymptotic solutions of nonlinear second order differential equations with variable coefficients. *J. Appl. Math. Mech.*, **23**, 515–526. [176]

Lagerstrom, P. A. (1957). Note on the preceding two papers. *J. Math. Mech.*, **6**, 605–606. [37]

Lagerstrom, P. A. (1961). Méthodes asymptotiques pour l'étude des equations de Navier-Stokes. *Lecture Notes*, Institut Henri Poincaré: Paris. (Translated by T. J. Tyson, California Institute of Technology, Pasadena, CA, 1965.) [29,45,104]

Lagerstrom, P. A. (1964). Laminar flow theory. *High Speed Aerodynamics and Jet Propulsion* (F. K. Moore, ed.), **4**, 20–285, Princeton University Press, Princeton, NJ. [21,175,216,218,230,235]

Lagerstrom, P. A. (1975). Solutions of the Navier-Stokes equations at large Reynolds number. *SIAM J. Appl. Math.*, **28**, 202–214. [176,216]

Lagerstrom, P. A. (1976). Forms of singular asymptotic expansions in layer-type problems. *Rocky Mountain Journal of Mathematics*, **6** No. 4, 609–635. [78,86,98,99,101]

Lagerstrom, P. A., and Casten, R. G. (1972). Basic concepts underlying singular perturbation techniques. *SIAM Rev.*, **14**, 63–120. [31,44,107,125]

Lagerstrom, P. A., and Cole, J. D. (1955). Examples illustrating expansion procedures for the Navier-Stokes equations. *J. Rational Mech. Anal.*, **4**, 817–882. [4,111,193,194,225]

Lagerstrom, P. A., Cole, J. D., and Trilling, L. (1949). Problems in the theory of viscous compressible fluids. Caltech, Pasadena. [175,193,210]

Lagerstrom, P. A., and Kevorkian, J. (1963a). Matched-conic approximation to the two fixed force-center problem. *The Astronomical Journal*, **68** March 1963, 84–92. [151,153,157]

Lagerstrom, P. A., and Kevorkian, J. (1963b). Earth-to-Moon trajectories in the restricted three-body problem. *Journal de Mécanique*, **2** June 1963, 129–218. (Reprinted in *Astrodynamics and Celestial Mechanics*, Vol. XII, June 1971.) [151,157]

Lagerstrom, P. A., and Kevorkian, J. (1963c). Earth-to-Moon trajectories with minimal energy. *Journal de Mécanique*, **2** December 1963, 493–504. [157]

Lagerstrom, P. A., and Reinelt, D. A. (1984). Note on logarithmic switchback terms in regular and singular perturbation expansions. *SIAM J. Appl. Math.*, **44** (3), 451–562. [72,107,119,123]

Lamb, H. (1911). On the uniform motion of a sphere through a viscous fluid. *Phil. Mag*, **21**, 112–121. [224]

Landau, L. D., and Lifshitz, E. M. (1959). *Fluid Mechanics*. Translated from Russian by Sykes and Reid, Addison-Wesley, Reading, MA. [135]

Latta, G. E. (1951). Singular perturbation problems. Doctoral thesis, California Institute of Technology, Pasadena. [175,196]

Legner, H. (1971). On optimal coordinates in boundary-layer theory. Doctoral thesis, Stanford University, Stanford, CA. [193]

Lemke, H. (1913). Über die differentialgleichungen, welche den Gleichgewichtszustand eines gasförmigen Himmelskorpers bestimmen, dessen teile gegeneinander nach dem Newtonschen gesetze gravitieren. *J. für die Reine und Angewandte Mathematik*, **142** 118–145. [128]

Levinson, N. (1950). The first boundary value problem for $\epsilon \Delta u + A(x,y)u_x + B(x,y)u_y + C(x,y)u = D(x,y)$ for small ϵ. *Ann. Math.*, **51**, 428–445. [175,177]

Lewis, J., and Carrier, G. F. (1949). Some remarks on the flat plate boundary layer. *Q. Appl. Math.*, **7**, 228–234. [234]

Lin, C. C., and Segel, L. A. (1974). *Mathematics Applied to Deterministic Problems in the Natural Sciences*. Macmillan, New York. [13]

Lo, Lilian (1983). The meniscus on a needle — a lesson in matching. *J. Fluid Mech.*, **132**, 65–78. [145,148]

Applied Mathematical Sciences

cont. from page ii